Lecture Notes in Mathematics

Edited by A. Dold and B. Eckmann

979

Mathematical Theories of Optimization

Proceedings of the International Conference
Held in S. Margherita Ligure (Genova)
November 30 – December 4, 1981

Edited by J.P. Cecconi and T. Zolezzi

Springer-Verlag
Berlin Heidelberg New York Tokyo 1983

Editors

Jaurés P. Cecconi
Tullio Zolezzi
Istituto per la Matematica Applicata CNR
Presso Istituto Matematico, Università di Genova
Via L.B. Alberti 4, Genova, Italia

AMS Subject Classifications (1980): 49-XX, 35-XX, 93-XX

ISBN 3-540-11999-X Springer-Verlag Berlin Heidelberg New York Tokyo
ISBN 0-387-11999-X Springer-Verlag New York Heidelberg Berlin Tokyo

Library of Congress Cataloging in Publication Data. Main entry under title: Mathematical theories of optimization. (Lecture notes in mathematics; 979) 1. Mathematical optimization–Congresses. 2. Calculus of variations–Congresses. 3. Differential equations, Partial–Congresses. I. Cecconi, Jaures P. II. Zolezzi, T. (Tullio), 1942. III. Series: Lecture notes in mathematics (Springer-Verlag); 979. QA3.L28 no. 979 [QA402.5] 510s [515] 83-588 ISBN 0-387-11999-X (U.S.)

Printing and binding: Beltz Offsetdruck, Hemsbach/Bergstr.
2146/3140-543210

This volume contains the lectures and contributed papers presented at the international conference on "Mathematical Theories of Optimization" held in S. Margherita Ligure (Genova) from November 30th to December 4th, 1981, and organized by the Istituto per la Matematica Applicata del C.N.R., Genova.

Most of the papers were presented at this conference and a few others were contributed by mathematicians unable to attend.

The conference aimed at up-to-date contributions and surveys in mathematical optimizations. The research papers of these proceedings cover many subjects, from optimal control to calculus of variations, from operations research to non smooth optimization and Gamma Convergence.

The meeting brought together many mathematicians from many countries who represented a wide range of interests in the subject, and was attended by approximately eighty mathematicians.

We use this opportunity to thank the members of the Istituto per la Matematica del C.N.R. di Genova for their support in the organization. We wish to express our thanks to the institutions whose financial support made the conference possible. These are Consiglio Nazionale delle Ricerche, Istituto di Matematica della Università di Genova and Regione Liguria.

Last but not least we wish to thank the editors of the Lecture Notes for their help.

J. P. Cecconi

T. Zolezzi

TABLE OF CONTENTS

A CONVERGENCE FOR BIVARIATE FUNCTIONS AIMED AT THE CONVERGENCE OF SADDLE VALUES

Hedy Attouch
Department of Mathematics
University of Orsay-Paris
France

Roger J.-B. Wets*
Department of Mathematics
University of Kentucky
USA

ABSTRACT

Epi/hypo-convergence is introduced from a variational viewpoint. The known topological properties are reviewed and extended. Finally, it is shown that the (partial) Legendre-Fenchel transform is bicontinuous with respect to the topology induced by epi/hypo-convergence on the space of convex-concave bivariate functions.

*Partially supported by a Guggenheim Fellowship.

1. INTRODUCTION.

One of our motivation is to introduce a notion of convergence well adapted to the study of extremal problems that can not be reduced to minimization problems. For example, let us consider a sequence of variational inequalities

$$(1.1_\varepsilon)\qquad \begin{cases} \langle A^\varepsilon u_\varepsilon - f, v - u_\varepsilon \rangle \geq 0 & \forall v \in K^\varepsilon \\ u_\varepsilon \in K^\varepsilon \end{cases}$$

where ε is a parameter describing an approximation, or a perturbation, homogenization... procedure. The operators $(A^\varepsilon)_{\varepsilon>0}$, the contraints K^ε are varying with ε, and the problem is to determine the behaviour, as ε goes to zero, of the solutions $(u_\varepsilon)_{\varepsilon>0}$ of the corresponding problems (1_ε). When the operators A^ε are subdifferentials of convex functionals and K^ε is convex, the problems (1_ε) can be viewed as minimization ones; but in general (take A^ε general operators of the calculus of variations, for example non symmetric second order elliptic operators, parabolic operators...) (1_ε) does not come from a minimization problem. However, it can always be expressed as a saddle value problem, under rather general assumptions, as already noticed by Glowinski, Lions and Tremolières [1], see also Rockafellar [13].

1.2 PROPOSITION. *Let* V *be a vector space and denote by* V' *its dual space. Given* $A : V \longrightarrow V'$, *a monotone operator, i.e. for all* $x, y \in V$, $\langle Ax - Ay, x - y \rangle \geq 0$, *and* $\phi : V \longrightarrow]-\infty, +\infty]$ *a real-valued function defined on* V, $\phi \not\equiv \infty$, *for any* $f \in V'$, *the following statements are equivalent :*

(i) u *is a solution of the variational inequality*

$$(1.3)\qquad \langle Au - f, v - u \rangle + \phi(v) - \phi(u) \geq 0 \qquad \forall v \in V$$

(ii) (u,u) *is a saddle point of the function* $H : V \times V \longrightarrow \bar{\mathbb{R}}$

$$H(u,v) = \langle Au - f, u - v \rangle + \phi(u) - \phi(v).$$

PROOF. By definition of H, u is a solution of the variational inequality (1.3), if and only if

(1.4) $\quad H(u,v) \leq 0, \quad$ for all $v \in V$.

Note that $H(u,v) = 0$ whenever u is a solution of (1.3). Thus it necessarily satisfies

$$H(u,v) \leq H(u,u) \qquad \text{for all } v \in V.$$

On the other hand, for all $w \in V$

$$\begin{aligned} H(w,u) &= \langle Aw - f, w - u \rangle + \phi(w) - \phi(u) \\ &= \langle Aw - Au, w - u \rangle + \langle Au - f, w - u \rangle + \phi(w) - \phi(u) \\ &= \langle Aw - Au, w - u \rangle - H(u,w) \\ &\geq 0. \end{aligned}$$

This last inequality following from the monotonicity of A and (1.4). So, for all $v \in V$ and $w \in V$, $H(u,v) \leq H(u,u) \leq H(w,u)$ which means that (u,u) is a saddle point of H.

Conversely if u is a saddle point of H, for all $v \in V$

$$H(u,v) \leq H(u,u) = 0,$$

which from (1.4) implies that u is a solution of the variational inequality (1.3).□

Let us now examine an important example : take $V = H_0^1(\Omega)$, Ω a bounded regular open set in $\mathbb{R}^N$, $V' = H^{-1}(\Omega)$.

$$A^\varepsilon(u) = -\sum_{i,j=1}^{N} \frac{\partial}{\partial x_i}\left(a_{ij}^\varepsilon(x) \frac{\partial u}{\partial x_j}\right)$$

where the $a_{ij}^\varepsilon \in L^\infty(\Omega)$ satisfy :

$$|a^{\varepsilon}_{ij}| \leq M$$

$$\sum a^{\varepsilon}_{ij}\, \xi_i\, \xi_j \geq \lambda_o |\xi|^2$$

with $\lambda_o > 0$ and M independent of x and ε. We do *not* require that the matrix (a^{ε}_{ij}) be symmetric, i.e. a^{ε}_{ij} is not necessarily equal to a^{ε}_{ji}. This class of problems is being studied by A. Brillard. For simplicity, we only consider the case with no constraints on u, i.e. $K^{\varepsilon} = V$ or equivalently $\phi^{\varepsilon} \equiv 0$. So, the variational inequalities (1.1_{ε}) reduce to the linear partial differential equations $A^{\varepsilon}u = f$. The natural notion of convergence $A^{\varepsilon} \xrightarrow{G} A$, as introduced by De Giorgi and Spagnolo [2] and Murat and Tartar [3], is

(1.5) $\quad$ for all $f \in H^{-1}(\Omega)$: $u_{\varepsilon} = (A^{\varepsilon})^{-1} f \xrightarrow[w-V]{} u = (A)^{-1} f$,

i.e. for the weak topology of $H^1_o(\Omega)$. Let us examine what is the corresponding notion of convergence for the saddle-functions

(1.6) $\quad H^{\varepsilon}(u,v) = \langle A^{\varepsilon}u,\, u - v \rangle.$

1.7. PROPOSITION. *The following statements are equivalent :*

(i) $\quad A^{\varepsilon} \xrightarrow{G} A$

(ii) $\quad H^{\varepsilon} \longrightarrow H$ *in the following sense : for every* $u,v \in V$

(1.8) $$\left| \begin{array}{l} \forall\, u_{\varepsilon} \rightharpoonup u \;\exists\, v_{\varepsilon} \rightharpoonup v \text{ such that } \liminf\limits_{\varepsilon\to 0} H^{\varepsilon}(u_{\varepsilon},v_{\varepsilon}) \geq H(u,v), \\ \forall\, v_{\varepsilon} \rightharpoonup v \;\exists\, u_{\varepsilon} \rightharpoonup u \text{ such that } H(u,v) \geq \limsup\limits_{\varepsilon\to 0} H^{\varepsilon}(u_{\varepsilon},v_{\varepsilon}). \end{array} \right.$$

where $\rightharpoonup$ *denotes weak-convergence.*

PROOF. Let us first note that $A^{\varepsilon} \xrightarrow{G} A$ if and only if $(A^{\varepsilon})^t \xrightarrow{G} A^t$ where $(A^{\varepsilon})^t$ and A^t are the elliptic operators with the transposed matrix $(a^{\varepsilon}_{ij})^t = a^{\varepsilon}_{ji}$ and $(a_{ij})^t = a_{ji}$.

Let us first verify that (i) $\Rightarrow$ (ii). Fix $u_\varepsilon \longrightarrow u$ and $v \in V$. We are looking for a sequence $v_\varepsilon \longrightarrow v$ such that

$$\liminf_{\varepsilon \to 0} \langle A^\varepsilon u_\varepsilon, u_\varepsilon - v_\varepsilon \rangle \geq \langle Au, u - v \rangle$$

Let w_ε be the solution of

$$(A^\varepsilon)^t w_\varepsilon = A^t(u - v). \tag{1.9}$$

By the definition of G-convergence for the sequence of operators $(A^\varepsilon)^t$ to A^t, as $\varepsilon \downarrow 0$ we have

$$w_\varepsilon \longrightarrow u - v$$

in the weak topology of V. Set

$$v_\varepsilon = u_\varepsilon - w_\varepsilon .$$

Then $v_\varepsilon \longrightarrow u - (u - v) = v$ and $u_\varepsilon - v_\varepsilon = w_\varepsilon$. Hence

$$\begin{aligned} \langle A^\varepsilon u_\varepsilon, u_\varepsilon - v_\varepsilon \rangle &= \langle A^\varepsilon u_\varepsilon, w_\varepsilon \rangle \\ &= \langle u_\varepsilon, (A^\varepsilon)^t w_\varepsilon \rangle \\ &= \langle u_\varepsilon, A^t(u - v) \rangle \end{aligned}$$

as follows from (1.9). Letting ε tend to 0, we get

$$\begin{aligned} \lim_{\varepsilon \to 0} \langle A^\varepsilon u_\varepsilon, u_\varepsilon - v_\varepsilon \rangle &= \langle u, A^t(u - v) \rangle \\ &= \langle Au, u - v \rangle. \end{aligned}$$

This completes the proof of the first part of (1.8). Next, fix $v_\varepsilon \longrightarrow v$ and $\bar{u} \in V$. This time we search for a sequence $u_\varepsilon \longrightarrow \bar{u}$ such that

$$\langle A\bar{u}, \bar{u} - v \rangle \geq \limsup_{\varepsilon \to 0} \langle A^\varepsilon u_\varepsilon, u_\varepsilon - v_\varepsilon \rangle.$$

Let u_ε be the solution of the equation $A^\varepsilon u = A\bar{u}$. Then

$$\langle A^\varepsilon u_\varepsilon, u_\varepsilon - v_\varepsilon \rangle = \langle A\bar{u}, u_\varepsilon - v_\varepsilon \rangle$$

and since $u_\varepsilon \longrightarrow \bar{u}$ and $v_\varepsilon \longrightarrow v$ we get

$$\lim_{\varepsilon \to 0} \langle A^\varepsilon u_\varepsilon, u_\varepsilon - v_\varepsilon \rangle = \langle A\bar{u}, \bar{u} - v \rangle$$

Next we prove that (ii) $\Rightarrow$ (i), that is to say, we verify if the convergence of the saddle functions H_n has the desired variational properties. Fix $f \in V$ and for $\varepsilon > 0$, let u_ε denote the solution of the equation $A^\varepsilon u = f$. The uniform coerciveness of the operators A^ε yields the boundedness of the u^ε in V. Passing to a subsequence if necessary, we have that

$$u_\varepsilon \longrightarrow \bar{u},$$

for some $\bar{u}$. To complete the proof we need to show that $A\bar{u} = f$. This will follow from the uniqueness of the solution of the equation $Au = f$. From (1.8), for any $v \in V$ there exists $v_\varepsilon \longrightarrow v$ such that

$$\liminf_{\varepsilon \to 0} \langle A^\varepsilon u_\varepsilon, u_\varepsilon - v_\varepsilon \rangle \geq \langle A\bar{u}, \bar{u} - v \rangle$$

which means that

$$\liminf_{\varepsilon \to 0} \langle f, u_\varepsilon - v_\varepsilon \rangle \geq \langle A\bar{u}, \bar{u} - v \rangle$$

or still

$$\langle f, \bar{u} - v \rangle \geq \langle A\bar{u}, \bar{u} - v \rangle,$$

and thus for all $v \in V$

$$\langle A\bar{u} - f, \bar{u} - v \rangle \leq 0$$

and $A\bar{u} = f$. □

In the preceeding example, we like to stress the fact that the saddle functions H^ε are not convex-concave. The lack of convexity comes from the non-symmetry of the monotone operators A^ε. Note also that in this example is not quite necessary to require both parts of (1.8), since the first part implies the second. This will not be the case in general, both conditions of (1.8) are usually necessary to obtain the desired variational properties.

Our next example is intented to illustrate the problems that arise in connection with Lagrangians and Hamiltonians. Let us consider the following class of optimization problems, for $\nu = 1, 2, \ldots$

(1.10$_\nu$) Minimize $f_o^\nu(x)$

subject to $f_i^\nu(x) \leq 0$ $\quad i = 1,\dots,m$

$x \in C \subset X$

with X a reflexive Banach space and C a closed subset. The associated Lagrangian function is

$$L_\nu(x,y) = \begin{cases} f_o^\nu(x) + \sum_{i=1}^m y_i f_i^\nu(x) & \text{if } x \in C \text{ and } y \geq 0 \\ +\infty \text{ if } x \notin C \text{ and } y \geq 0 \\ -\infty \text{ otherwise.} \end{cases} \tag{1.11}$$

We think of the problems (1.10$_\nu$) and their Lagrangians as the approximates of some limit problem :

(1.12) Minimize $f_o(x)$

subject to $f_i(x) \leq 0$ $\quad i = 1,\dots,m$

$x \in C \subset X$

with associated Lagrangian

$$L(x,y) = \begin{cases} f_o(x) + \sum_{i=1}^m y_i f_i(x) & \text{if } x \in C \text{ and } y \geq 0 \\ +\infty \text{ if } x \notin C \text{ and } y \geq 0 \\ -\infty \text{ otherwise.} \end{cases} \tag{1.13}$$

A typical situation is when the problems (1.10$_\nu$) are obtained from (1.12) as the result of penalization or barrier terms being added to the objective, or when the (1.10$_\nu$) are the restrictions of (1.12) to finite dimensional subspaces of X, and so on. In particular, when dealing with numerical procedures, one is naturally interested in the convergence of the solutions, but also in the convergence of the multipliers, for reason of stability [4] or to be able to calculate rates of convergence such as in augmented Lagrangian methods. From the convergence of the $\{f_i^\nu, \nu = 1,\dots\}$ to the f_i

one cannot conclude in general that the feasible sets

$$S_\nu = \{x \in C | f_i^\nu(x) \le 0, \quad i = 1,\dots,m\}$$

converge to the feasible set of the limit problem,

$$S = \{x \in C | f_i(x) \le 0, \quad i = 1,\dots,m\}.$$

A fortiori, it is not possible to obtain the convergence of the infima or of the optimal solutions. However, there are some relatively weak conditions that can be imposed on the convergence of the objectives and of the constraints that will guarantee the convergence of the Lagrangians L_ν to L in a sense similar to that induced by G-convergence on the saddle functions (1.6) associated with the partial differential equations $A^\varepsilon u = f$. The sought for, convergence of the solutions *and* multipliers will ensue.

Given $\{f; f^\nu : X \longrightarrow \overline{R}, \nu = 1,\dots\}$ a collection of functions, we say that the f^ν *epi-convergence* to f if for all x

(1.14) $$\text{for all } x_\nu \longrightarrow x, \quad \liminf_{\nu\to\infty} f^\nu(x\;) \ge f(x),$$

and

(1.15) $$\text{there exists } x_\nu \longrightarrow x \text{ with } \limsup_{\nu\to\infty} f^\nu(x_\nu) \le f(x).$$

As is well-known, epi-convergence is neither implied nor does it imply pointwise convergence, but they coincide, for example, if the sequence of functions is monotone, either increasing or decreasing (provided f is lower semicontinuous). We have so-called *continuous convergence* if condition (1.15) is replaced by the stronger requirement

(1.16) $$\text{for all } x_\nu \longrightarrow x, \limsup_{\nu\to\infty} f^\nu(x_\nu) \le f(x).$$

Continuous convergence is much stronger that both epi- and pointwise-convergence.

1.17 PROPOSITION. *Suppose the* $\{f_o^\nu, \nu = 1,\ldots\}$ *epi-converge* f_o, *and for all* $i = 1,\ldots,m$, *the* $\{f_i^\nu, \nu = 1,\ldots\}$ *continuously converge to* f_i. *Then, the associated Lagrangian functions* L_ν *converge to the Lagrangian* L *in the following sense : for all* $x \in X$ *and* $y \in Y$

$$(1.18)\quad \left|\begin{array}{l} \textit{for any } x_\nu \longrightarrow x\,,\ \textit{there exists } y_\nu \longrightarrow y \ \textit{ such that} \\ \qquad\qquad\qquad \liminf_{\nu\to\infty} L_\nu(x_\nu,y_\nu) \geq L(x,y) \\ \textit{for any } y_\nu \longrightarrow y\,,\ \textit{there exists } x_\nu \longrightarrow x \ \textit{ such that} \\ \qquad\qquad\qquad \limsup_{\nu\to\infty} L_\nu(x_\nu,y_\nu) \leq L(x,y). \end{array}\right.$$

Moreover, suppose that the Lagrangians L_ν *converge to* L *in the above sense, and for some subsequence* $\{\nu_k, k = 1,\ldots\}$ *the sequence* $\{(\bar{x}^k,\bar{y}^k), k = 1,\ldots\}$, *which converge to* $(\bar{x},\bar{y})$ *is such that* $\bar{x}^k$ *solves problem* (1.10_{ν_k}) *and* $\bar{y}^k$ *is a (Lagrange) multiplier. Then* $(\bar{x},\bar{y})$ *is a saddle point of* L. *And if* L *is convex-concave, then* $\bar{x}$ *solves* (1.12) *and* $\bar{y}$ *is an associated multiplier.*

PROOF. We start by showing that the conditions imposed on the f_o^ν and $\{f_i^\nu, i = 1,\ldots,m\}$ yield (1.18). Let x^ν be any sequence converging to x and set $y^\nu = y$ for all ν. We have to verify that when $x \in C$ and $y \geq 0$

$$\liminf_{\nu\to\infty} \Big(f_o^\nu(x_\nu) + \sum_{i=1}^m y_i\, f_i(x_\nu)\Big) \geq f_o(x) + \sum_{i=1}^m y_i\, f_i(x),$$

the cases when $y \ngeq 0$ and/or $x \notin C$ are automatically satisfied. Since C is closed, any sequence that converges to $x \notin C$ is such that $x^\nu \in X\setminus C$ for ν sufficiently large. The inequality in fact follows directly from (1.14) which is satisfied by both the epi-convergence of the f_o^ν and the continuous convergence of the f_i^ν, $i = 1,\ldots,m$.

Next we have to verify that for any sequence $y_\nu \longrightarrow y$, there exists $x^\nu \longrightarrow x$ such that when $x \in C$ and $y \geq 0$

$$\limsup_{\nu\to\infty} \left(f_o^\nu(x_\nu) + \sum_{i=1}^m y_i^\nu f_i^\nu(x_\nu)\right) \le f_o(x) + \sum_{i=1}^m y_i f_i(x).$$

When $x \notin C$ or/and $y \ngeq 0$ the desired relation between $\limsup_{\nu\to+\infty} L_\nu$ and L is automatically satisfied. The preceeding inequality then follows from (1.15) and (1.16).

If $\bar{x}^k$ solves (1.10_{ν_k}) and $\bar{y}^k$ is an associated multiplier, we have that for $i = 1,\dots,m$

$$\bar{y}^k \ge 0 \ , \ f_i^k(\bar{x}^k) \le 0 \quad \text{and } \bar{y}_i^k f_i^{\nu_k}(\bar{x}^k) = 0,$$

and

$$\bar{x}^k \in \operatorname{argmin}_{x \in C} \left(f_o^{\nu_k}(x) + \sum_{i=1}^m y_i^k f_i^{\nu_k}(x)\right) .$$

This is equivalent to : for all x and y

$$L_{\nu_k}(\bar{x}^k, y) \le L_{\nu_k}(\bar{x}^k, \bar{y}^k) \le L_{\nu_k}(x, \bar{y}^k),$$

with the first inequality equivalent to the first part of the optimality conditions and the second inequality is just a restatement of the second part of the optimality conditions.

Thus the assertion will be complete if we show that $(\bar{x},\bar{y}) = \lim_{k\to\infty}(\bar{x}_k,\bar{y}_k)$ is a saddle point of L, i.e.

$$L(\bar{x},y) \le L(\bar{x},\bar{y}) \le L(x,\bar{y}).$$

First note that if the sequence L_ν converges to L in the sense of (1.8) so does the subsequence $\{L_{\nu_k}, k = 1,\dots\}$. Since the $(\bar{x}_k,\bar{y}_k)$ are saddle points, for any pair of sequences $\{x^k, k = 1\dots\}$ and $\{y^k, k = 1\dots\}$ converging to x and y respectively, we have

$$\liminf_{k\to\infty} L_{\nu_k}(\bar{x}^k, y^k) \le \liminf_{k\to\infty} L_{\nu_k}(\bar{x}^k, \bar{y}^k)$$

$$\le \limsup_{k\to\infty} L_{\nu_k}(\bar{x}^k, \bar{y}^k) \le \limsup_{k\to\infty} L_{\nu_k}(x^k, \bar{y}^k)$$

In particular the $\{x^k, k = 1\dots\}$ and $\{y^k, k = 1\dots\}$ could have been those satisfying (1.18), and hence

$$L(\bar{x},y) \le L(x,\bar{y})$$

which yields the saddle point property of $(\bar{x},\bar{y})$. This in turn yields the final assertions of the Proposition. □

Proposition 1.17 extends the results of T. Zolezzi [5, Theorem 4] about stability in mathematical programming. Many assumptions, such as compactness conditions on the feasible regions, can be ignored when one use this type of convergence rather than convergence notions that only involve the x variables.

2. EPI/HYPO-CONVERGENCE FROM A VARIATIONAL VIEWPOINT.

Let $\{F^\nu : X \times Y \longrightarrow \bar{R} = [-\infty, +\infty], \nu = 1,\dots\}$ be a sequence of bivariate functions, and for each ν, let (x_ν, y_ν) denote a saddle point of F^ν, i.e.

(2.1) $$F^\nu(x_\nu, y) \le F^\nu(x_\nu, y_\nu) \le F^\nu(x, y_\nu) \quad \text{for all } x \in X \text{ and } y \in Y.$$

We show that the convergence of saddle points and saddle values implicitly subsumes certain topological properties for the sequence $\{F^\nu, \nu = 1,\dots\}$ which lead naturally to the definition of epi/hypo-convergence.

Relation (2.1) yields estimates for x_ν and y_ν and hence also relative compactness properties for the sequence $\{(x_\nu, y_\nu), \nu = 1\dots\}$. Let us assume that for some topologies τ and σ, a subsequence $\{x_{\nu_k}, k = 1\dots\}$ τ-converge to $\bar{x}$ and $\{y_{\nu_k}, k = 1\dots\}$ σ-converge to $\bar{y} \in Y$. Neither τ nor σ need be given a priori, they could for example, be the result of some uniform coerciveness properties of the F^ν and compact embeddings. For any pair $(x,y) \in X \times Y$, not only does (2.1) hold but also

$$\sup_{v \in V} F^\nu(x_\nu, v) \le \inf_{u \in U} F^\nu(u, y_\nu)$$

for all $U \in \mathcal{N}_\tau(x)$ and $V \in \mathcal{N}_\sigma(x)$ where $\mathcal{N}_\tau(x)$ and $\mathcal{N}_\sigma(y)$ are the τ- and σ-neighborhood systems of x and y respectively. Since $x_{\nu_k} \xrightarrow[\tau]{} \bar{x}$ and $y_{\nu_k} \xrightarrow[\sigma]{} \bar{y}$, for any pair $(U_{\bar{x}}, V_{\bar{y}}) \in \mathcal{N}_\tau(\bar{x}) \times \mathcal{N}_\sigma(\bar{y})$ and k large enough

$$x_{\nu_k} \in U_{\bar{x}} \quad \text{and} \quad y_{\nu_k} \in V_{\bar{y}}$$

and hence

(2.2) $$\inf_{u \in U_{\bar{x}}} \sup_{v \in V} F_{\nu_k}(u,v) \le \sup_{v \in V_{\bar{y}}} \inf_{u \in U} F_{\nu_k}(u,v).$$

This holds for any convergent subsequence of the $\{(x_\nu,y_\nu),\nu = 1,\ldots\}$ and since for any sequence of extended real-numbers $\{a_\nu,\nu = 1,\ldots\}$

$$\inf_{\{\nu_k\}\subset\{1,\ldots\}} \liminf_{k\to\infty} a_{\nu_k} = \liminf_{\nu\to\infty} a_\nu$$

and

$$\sup_{\{\nu_k\}\subset\{1,\ldots\}} \limsup_{k\to\infty} a_{\nu_k} = \limsup_{\nu\to\infty} a_\nu$$

it follows that

$$\begin{aligned}&\liminf_{\nu\to\infty} \inf_{u\in U_{\bar{x}}} \sup_{v\in V} F^\nu(u,v)\\ \text{(2.3)}\qquad &\qquad\le \limsup_{\nu\to\infty} \sup_{v\in V_{\bar{y}}} \inf_{u\in U} F^\nu(u,v),\end{aligned}$$

which must hold for any pair $(\bar{x},\bar{y})$.

To extract as much information from (2.3) at the (local) pointwise level, we use the fact that the above holds for all $U\in\mathcal{N}_\tau(x)$, $U_{\bar{x}}\in\mathcal{N}_\tau(\bar{x})$, $V\in\mathcal{N}_\sigma(y)$ and $V_{\bar{y}}\in\mathcal{N}_\sigma(\bar{y})$ to take infs and sups with respect to these neighborhood systems. Since inf sup $\ge$ sup inf, and because the lim inf and lim sup that appear in (2.3) are monotone with respect to U and V as they decrease to x and y respectively, the sharpest inequality one can obtain at x and y is

$$\begin{aligned}&\inf_{V\in\mathcal{N}_\sigma(y)} \sup_{U\in\mathcal{N}_\tau(\bar{x})} \liminf_{\nu\to\infty} \inf_{u\in U} \sup_{v\in V} F^\nu(u,v)\\ \text{(2.4)}\qquad &\le \sup_{U\in\mathcal{N}_\tau(x)} \inf_{V\in\mathcal{N}_\sigma(\bar{y})} \limsup_{\nu\to\infty} \sup_{v\in V} \inf_{u\in U} F^\nu(u,v).\end{aligned}$$

The expression which appears on the left of the inequality is a function of $\bar{x}$ and y, the one on the right depends on x and $\bar{y}$. Let us denote them by h/e-li F^ν and e/h-ls F^ν respectively ; this notation to be justified later on. Rewriting (2.4), we see that whenever $\bar{x}$ and $\bar{y}$ are limit points of saddle points, then

(2.5) $$\text{h/e-li } F^\nu(\bar{x},y) \le \text{e/h-ls } F^\nu(x,\bar{y})$$

for all $x \in X$ and $y \in Y$. In particular this implies that

$$h/e\text{-li } F^{\nu}(\bar{x},y) \le e/h\text{-ls } F^{\nu}(\bar{x},\bar{y}) \text{ for all } y$$

and

$$h/e\text{-li } F^{\nu}(\bar{x},\bar{y}) \le e/h\text{-ls } F^{\nu}(x,\bar{y}) \text{ for all } x.$$

Suppose $F' = h/e\text{-li } F^{\nu} = e/h\text{-ls } F^{\nu}$, then the preceeding inequalities imply that $(\bar{x},\bar{y})$ is a saddle point of F'. Since admittedly we seek a notion of convergence for bivariate functions that will yield the convergence of the saddle points to a saddle point of the limit function, the function F', if it exists, is a natural candidate. This is somewhat too restrictive and would exclude a large class of inter sting applications. In fact any function F with the property that

$$e/h\text{-ls } F^{\nu} \le F \le h/e\text{-li } F^{\nu} \tag{2.6}$$

will have the desired property, since then

$$F(\bar{x},y) \le h/e\text{-li } F^{\nu}(\bar{x},y) \le e/h\text{-ls } F^{\nu}(\bar{x},\bar{y}) \le F(\bar{x},\bar{y})$$

and

$$F(\bar{x},\bar{y}) \le h/e\text{-li } F^{\nu}(\bar{x},\bar{y}) \le e/h\text{-ls } F^{\nu}(x,\bar{y}) \le F(x,\bar{y})$$

for all $x \in X$ and $y \in Y$, i.e. $(\bar{x},\bar{y})$ is a saddle point of F.

We started with a collection of bivariate functions whose only property was to possess a (sub)sequence of convergent saddle points. If the limit of such a sequence is to be a saddle point of the limit function, we are led to certain conditions that must be satisfied by the limit function(s), and it is precisely these conditions that we shall use for the definition of epi/hypo-convergence.

We now review this at a somewhat more formal level. As we have seen, we need the two functions associated to the sequence $\{F^{\nu}, \nu = 1,\dots\}$

$$h/e\text{-li } F^{\nu} = h_{\sigma}/e_{\tau}\text{-li } F^{\nu} = \text{hypo}_{\sigma}/\text{epi}_{\tau}\text{-}\liminf_{\nu\to\infty} F^{\nu}$$

$$e/h - \text{ls } F^{\nu} = e_{\tau}/h_{\sigma} - \text{ls } F^{\nu} = \text{epi}_{\tau}/\text{hypo}_{\sigma} - \limsup_{\nu\to\infty} F^{\nu}$$

with

$$h_\sigma/e_\tau\text{-li } F^\nu(x,y) = \tag{2.7}$$
$$\inf_{V \in \mathcal{N}_\sigma(y)} \sup_{U \in \mathcal{N}_\tau(x)} \lim_{\nu\to\infty}\inf \inf_{u \in U} \sup_{v \in V} F^\nu(u,v)$$

called the *hypo/epi-limit inferior*, and

$$e_\tau/h_\sigma\text{-ls } F^\nu(x,y) = \tag{2.8}$$
$$\sup_{U \in \mathcal{N}_\tau(x)} \inf_{V \in \mathcal{N}_\sigma(y)} \lim_{\nu\to\infty}\sup \sup_{v \in V} \inf_{u \in U} F^\nu(u,v)$$

called the *epi/hypo-limit superior*. The properties of these limit functions will be reviewed in the next Section.

A (bivariate) function F is said to be an *epi/hypo-limit* of the sequence $\{F^\nu, \nu = 1,\ldots\}$ if

$$e_\tau/h_\sigma\text{-ls } F^\nu \le F \le h_\sigma/e_\tau\text{-li } F^\nu. \tag{2.9}$$

Thus in general epi/hypo-limits are not unique, i.e. the topology induced by epi/hypo-convergence on the space of (bivariate) functions is not Hausdorff. This is intimately connected to the nature of saddle functions, as is again exemplified in Section 7.

As already suggested by our discussion, this is not the only type of convergence of bivariate functions that could be defined. In fact our two limit functions are just two among many possible limit functions introduced by De Giorgi [6] in a very general setting and called Γ-limits. In his notation

$$h_\sigma/e_\tau\text{-li } F^\nu(x,y) = \Gamma(N^-, \tau^-, \sigma^+) \lim_{\substack{\nu\to\infty \\ u\to x \\ v\to y}} F^\nu(u,v)$$

and

$$e_\tau/h_\sigma\text{-ls } F^\nu(x,y) = \Gamma(N^+, \sigma^+, \tau^-) \lim_{\substack{\nu\to\infty \\ v\to y \\ u\to x}} F^\nu(u,v)$$

(We have adopted a simplified notation because it carries important geometric information, cf. Section 3, that gets lost with the Γ-notation). It is however important to choose these two functions since, not only do they arise naturally from the convergence of saddle points, but in some sense they are the "minimal" pair, as made clear in Section 4 of [7] . Other definitions have been proposed by Cavazutti [8], [9], see also Sonntag [10], that imply epi/hypo-convergence, but unfortunately restrict somewhat the domain of applications.

Finally, observe that when the F^ν do not depend on y, then the definition of epi/hypo-convergence specializes to the classical definition of epi-convergence (with respect to the variable x). On the other hand if the F^ν do not depend on x, then epi/hypo-convergence is simply hypo-convergence. Thus, the theory contains both the theory of epi- and hypo-convergence.

The variational properties of epi/hypo-convergence, that motivated the definition, are formalized by the next Theorem.

2.10 THEOREM [7] . *Suppose* (X,τ) *and* (Y,σ) *are two topological spaces and* $\{F^\nu, \nu = 1,\ldots\}$ *a sequence of bivariate functions, defined on* $X\times Y$ *and with values in the extended reals, that* epi$_\tau$/hypo$_\sigma$-*converge to a function* F. *Suppose that for some subsequence of functions* $\{F_{\nu_k}, k = 1,\ldots\}$ *with saddle points* (x_k,y_k) i.e. *for all* $k = 1,\ldots$

$$F_{\nu_k}(x_k,y) \le F_{\nu_k}(x_k,y_k) \le F_{\nu_k}(x,y_k),$$

the saddle points converge with $\bar{x} = \tau\text{-}\lim_{k\to\infty} x_k$ *and* $\bar{y} = \sigma\text{-}\lim_{k\to\infty} y_k$. *Then* $(\bar{x},\bar{y})$ *is a saddle point of* F *and*

$$F(\bar{x},\bar{y}) = \lim_{k\to\infty} F_{\nu_k}(x_k,y_k)$$

The second property which gives to this notion of convergence a great flexibility and renders it significant, when applied to variational problems, is its stability properties with respect to a large class of perturbations.

2.11 THEOREM. *Suppose* X,Y *and the* $\{F^\nu, \nu = 1,\ldots\}$ *are as in Theorem 2.10 with*

$$F = \text{epi}_\tau/\text{hypo}_\sigma\text{-lim}_{\nu\to\infty} F^\nu.$$

Then, for any continuous function $G : (X,\tau)\times(Y,\sigma) \longrightarrow R$,

$$F + G = \text{epi}_\tau/\text{hypo}_\sigma\text{-lim}_{\nu\to\infty}(F^\nu + G).$$

PROOF. Since G is continuous at (x,y), for every $\varepsilon > 0$ there exists $U_\varepsilon \in \mathcal{N}_\tau(x)$ and $V_\varepsilon \in \mathcal{N}_\sigma(y)$ such that for all $u \in U_\varepsilon$, $v \in V_\varepsilon$

$$G(x,y) - \varepsilon \leq G(u,v) \leq G(x,y) + \varepsilon$$

From this, it follows that

$$e/h\text{-ls}(F^\nu + G)(x,y)$$

$$= \sup_{U\subset U_\varepsilon} \inf_{V\subset V_\varepsilon} \limsup_{\nu\to\infty} \sup_{v\in V} \inf_{u\in U}(F^\nu + G)(u,v)$$

$$\geq \sup_{U\subset U_\varepsilon} \inf_{V\subset V_\varepsilon} \limsup_{\nu\to\infty} \left[\sup_{v\in V} \inf_{u\in U}(F_\nu(u,v) + G(x,y) - \varepsilon)\right]$$

$$\geq (e/h\text{-ls } F^\nu)(x,y) + G(x,y) - \varepsilon.$$

This holds for every $\varepsilon > 0$ and thus

$$e/h\text{-ls}(F^\nu + G) \geq (e/h\text{-ls } F^\nu) + G.$$

Again using the continuity of G, one shows similarly the converse inequality which thus yields

$$e/h\text{-ls}(F^\nu + G) = G + e/h\text{-ls } F^\nu .$$

The same arguments can be used to obtain the identity involving $e/h\text{-li}(F^\nu + G)$ and $e/h\text{-li } F^\nu$. Thus, if

$$e/h\text{-ls } F^\nu \leq F \leq h/e\text{-li } F^\nu$$

it implies that

$$e/h\text{-ls}(F^\nu + G) \leq F + G \leq h/e\text{-li}(F^\nu + G)$$

which is precisely what is meant by $F+G = e/h\text{-lim}(F^\nu + G)$. □

3. PROPERTIES OF EPI/HYPO-LIMITS. GEOMETRICAL INTERPRETATION.

In general, an arbitrary collection of saddle functions does not have an epi/hypo-limit, and when it does the limit is not necessarily unique. This all comes from the fact that, *in general, the two limit functions are not comparable*. For example, let $X = Y = R$ and for ν odd

$$F^{\nu}(x,y) = \begin{cases} y\,x^{-1} \text{ on } [0,1]\times[0,1]\setminus\{(0,0)\}, \\ \text{arbitrary when } (x,y) = (0,0), \\ -\infty \text{ if } x\in[0,1] \text{ and } y\notin[0,1], \\ +\infty \text{ otherwise,} \end{cases}$$

and for ν even, $F^{\nu} = 2\,F_1$. Then

$$h/e\text{-li } F^{\nu}(x,y) = y\,x^{-1} < 2\,y\,x^{-1} = e/h\text{-ls } F^{\nu}(x,y) \quad \text{on }]0,1]\times]0,1]$$

but

$$h/e\text{-li } F^{\nu}(0,0) = +\infty > e/h\text{-ls } F^{\nu}(0,0) = 0.$$

When a sequence of bivariate functions $\{F^{\nu}, \nu = 1,\ldots\}$ epi-hypo-converges, its epi/hypo-limits form an interval

$$(3.1) \qquad [e/h\text{-ls } F^{\nu}, h/e\text{-li } F^{\nu}] = \{F: X\times Y \to \overline{R} \mid e/h\text{-ls } F^{\nu} \le F \le h/e\text{-li } F^{\nu}\}$$

These two limit functions have semicontinuity properties that follow directly from the definition and the following general lemma [7, Lemma 4.30] .

3.2 LEMMA. Suppose (X,τ) is a topological space and q an extended real valued function defined on the subsets of X. Then the function

$$x \longmapsto \sup_{U\in\mathcal{N}_{\tau}(x)} q(U)$$

is τ-lower semicontinuous, and the function

$$x \longmapsto \inf_{U\in\mathcal{N}_{\tau}(x)} q(U)$$

is τ-upper semicontinuous.

PROOF. Simply note that for every x

$$g(x) = \sup_{U\in \mathcal{N}_\tau(x)} q(U) \le cl_\tau g(x) = \sup_{U\in \mathcal{N}_\tau(x)} \inf_{u\in U} g(u),$$

as follows from the definition of g, since

$$q(U) \le \inf_{u\in U} g(u). \quad \Box$$

3.3. PROPOSITION. *Suppose* $\{F^\nu : (X,\sigma)\times(Y,\tau) \to \overline{R},\ \nu = 1,\ldots\}$ *is a sequence of bivariate functions. Then for all* y,

$$x \longmapsto e_\tau/h_\sigma\text{-ls } F^\nu(x,y)$$

is τ-*l.sc. in* x *, and for all* x

$$y \longmapsto h_\sigma/e_\tau\text{-li } F^\nu(x,y)$$

is σ-*u.sc. in* y.

One can also derive the semicontinuity properties of the limit functions from their geometrical interpretation as done in [7]

3.4. THEOREM. *Suppose* $\{F^\nu : (X,\sigma)\times(Y,\tau) \to \overline{R},\ \nu = 1,\ldots\}$ *is a sequence of bivariate functions. Then for every* $y\in Y$ *and* $x\in X$

$$\text{epi}(e/h\text{-ls } F^\nu)(\cdot,y) = \operatorname*{Lim\,inf}_{\substack{\nu\to\infty \\ y'\to y}} \text{epi } F^\nu(\cdot,y'),$$

and

$$\text{hypo}(h/e\text{-li } F^\nu)(x,\cdot) = \operatorname*{Lim\,inf}_{\substack{\nu\to\infty \\ x'\to x}} \text{hypo } F^\nu(x',\cdot).$$

Thus the epi-hypo-convergence of a sequence of bivariate functions is a limit concept that involves both epi- and hypo-convergence. That is clearly at the origin of our terminology. However note that both formulas require that limits be taken with respect to both ν and either x or y, and can not be equated with the epi- or hypo-convergence of the univariate functions $F^{\nu}(.,y)$ and $F^{\nu}(x,.)$ respectively. It is a much weaker notion, more sophisticated, which does not allow the two variables x and y to be handled independently.

4. EPI/HYPO-CONVERGENCE : THE METRIZABLE CASE.

In the metric case, or more generally when (X,τ) and (Y,σ) are metrizable, it is possible to give a representation of the limit functions in terms of sequences that turn out to be very useful in verifying epi/hypo-convergence, cf. [7, Corollary 4.4] . The formulas that we give here in terms of sequence--rather than subsequence-- are new and thus complement those given earlier in [7, Theorem 4.10 and Corollary 4.14] .

4.1 THEOREM. *Suppose* (X,τ) *and* (Y,σ) *are two metrizable spaces, and* $\{F^\nu : X\times Y \to \overline{R},\ \nu = 1,\ldots\}$ *a sequence of functions. Then for every* $(x,y) \in X\times Y$

$$(4.2)\qquad \text{e/h-ls } F^\nu(x,y) = \sup_{y_\nu \to_\sigma y}\ \min_{x_\nu \to_\tau x}\ \limsup_{\nu\to\infty} F^\nu(x_\nu,y_\nu),$$

$$= \sup_{\substack{\{\nu_k\}\subset N \\ y_k \to_\sigma y}}\ \min_{x_k \to_\tau x}\ \limsup_{k\to\infty} F^{\nu_k}(x_k,y_k),$$

and

$$(4.3)\qquad \text{h/e-li } F^\nu(x,y) = \inf_{x_\nu \to_\tau x}\ \max_{y_\nu \to_\sigma y}\ \liminf_{\nu\to\infty} F^\nu(x_\nu,y_\nu),$$

$$= \inf_{\substack{\{\nu_k\}\subset N \\ x_k \to_\tau x}}\ \max_{y_k \to_\sigma y}\ \liminf_{k\to\infty} F^{\nu_k}(x_k,y_k)$$

These characterizations of the limit functions yield directly the following criteria for epi/hypo-convergence.

4.4 COROLLARY. *Suppose* (X,τ) *and* (Y,σ) *are metrizable, and* $\{F^\nu : X\times Y \to \overline{R},\ \nu = 1,\ldots\}$ *a sequence of functions. Then the following assertions are equivalent*

(4.5) $$F = e_\tau/h_\sigma\text{-}\lim F^\nu$$

(4.6) (i) *For all* $y_\nu \underset{\sigma}{\rightarrow} y$, *there exists* $x_\nu \underset{\tau}{\rightarrow} x$ *such that*

$$\limsup_{\nu\to\infty} F^\nu(x_\nu,y_\nu) \le F(x,y),$$

and

(ii) *for all* $x_\nu \underset{\tau}{\rightarrow} x$, *there exists* $y_\nu \underset{\sigma}{\rightarrow} y$ *such that*

$$F(x,y) \le \liminf_{\nu\to\infty} F^\nu(x_\nu,y_\nu),$$

hold for all $(x,y) \in X\times Y$,

(4.7) (i) *for all* $\{\nu_k, k=1,\ldots\}\subset N$, $y_k \underset{\sigma}{\rightarrow} y$ *there exists* $x_k \underset{\tau}{\rightarrow} x$

such that $\limsup_{k\to\infty} F^{\nu_k}(x_k,y_k) \le F(x,y)$

and

(ii) *for all* $\{\nu_k\}\subset N$, $x_k \underset{\tau}{\rightarrow} x$ *there exists* $y_k \underset{\sigma}{\rightarrow} y$

such that $F(x,y) \le \liminf_{k\to\infty} F^{\nu_k}(x_k,y_k)$

hold for all $(x,y) \in X\times Y$.

PROOF OF THEOREM 4.1. Since $e/h\text{-ls}\, F^\nu = -(h/e\text{-li}(-F^\nu))$ it clearly suffices to prove one of the identities (4.2) or (4.3), say (4.3). We denote by G and H the following functions

$$G(x,y) = \inf_{x_\nu \underset{\tau}{\rightarrow} x} \sup_{y_\nu \underset{\sigma}{\rightarrow} y} \liminf_{\nu\to\infty} F^\nu(x_\nu,y_\nu),$$

and

$$H(x,y) = \inf_{\substack{\{\nu_k\}\subset N \\ x_k \underset{\tau}{\rightarrow} x}} \sup_{y_k \underset{\sigma}{\rightarrow} y} \liminf_{k\to\infty} F^{\nu_k}(x_k,y_k)$$

Obviously $G \ge H$, thus to obtain (4.3) we only need to prove that

$$G \le h/e\text{-li}\, F^\nu \le H.$$

First, we show that $G \le h/e\text{-li}\, F^\nu$. There is nothing to prove if $h/e\text{-li}\, F^\nu \equiv +\infty$, so let us assume that for some pair (x,y), $h/e\text{-li}\, F^\nu(x,y) < \infty$. Given any $\beta > h/e\text{-li}\, F^\nu(x,y)$, the definition of h/e-li F^ν yields a neighborhood $V_\beta \in \mathcal{N}_\sigma(y)$ such that for all $U \in \mathcal{N}_\tau(x)$

$$\beta \geq \liminf_{\nu\to\infty} \inf_{u \in U} \sup_{v \in V_\beta} F^\nu(u,v).$$

Let $\{U_\mu, \mu=1,\dots\}$ be a countable base of open neighborhoods of x, decreasing with μ to $\{x\}$. The preceeding inequality with U replaced by U_μ, implies the existence of a sequence $\{X_{\nu_\mu} \in U_\mu, \nu = 1,\dots\}$ such that

$$\beta \geq \liminf_{\nu\to\infty} \sup_{v \in V_\beta} F^\nu(x_{\nu_\mu}, v).$$

Since this holds for all μ, we get that

$$\beta \geq \limsup_{\mu\to\infty} \liminf_{\nu\to\infty} \sup_{v \in V_\beta} F^\nu(x_{\nu_\mu}, v).$$

We now rely on the Diagonalization Lemma, proved in the Appendix, to obtain a sequence $\{x_\nu = x_{\nu,\mu(\nu)}, \nu = 1,\dots\}$ with $\nu \longmapsto \mu(\nu)$ increasing (which implies that $x_\nu \xrightarrow[\tau]{} x$) such that

$$\beta \geq \liminf_{\nu\to\infty} \sup_{v \in V_\beta} F^\nu(x_\nu, v).$$

Now, for any sequence $y_\nu \xrightarrow[\sigma]{} y$, for ν sufficiently large $y_\nu \in V_\beta$ and hence

$$\beta \geq \liminf_{\nu\to\infty} F^\nu(x_\nu, y_\nu).$$

The above holds for any sequence $\{y_\nu, \nu = 1,\dots\}$ σ-converging to y. Using this and the fact that the x_ν τ-converge to x we have that

$$\beta \geq \sup_{y_\nu \xrightarrow[\sigma]{} y} \liminf_{\nu\to\infty} F^\nu(x_\nu, y_\nu)$$

and also

$$\beta \geq \inf_{x_\nu \xrightarrow[\tau]{} x} \sup_{y_\nu \xrightarrow[\sigma]{} y} \liminf_{\nu\to\infty} F^\nu(x_\nu, y_\nu) = G(x,y).$$

Since this holds for every $\beta <$ h/e-li $F^\nu(x,y)$ we get that h/e-li $F^\nu \geq G$.

Next we show that $H \geq$ h/e-li F^ν. Again there is nothing to prove if h/e-li $F^\nu \equiv -\infty$, so let us assume that for some (x,y), h/e-li $F^\nu(x,y) > -\infty$. The definition of h/e-li F^ν implies that given any $\alpha < \text{h/e-li}\, F^{\nu}(x,y)$ and any $V \in \mathcal{N}_{\sigma}(y)$ there corresponds a neighborhood $U = U_{\alpha,V}$ of x such that

$$\alpha < \liminf_{\nu\to\infty} \inf_{u\in U} \sup_{v\in V} F^{\nu}(u,v).$$

Let $\{V_{\mu},\ \mu=1,\ldots\}$ be a countable base of open neighborhoods of y, decreasing with μ to $\{y\}$. To any such V_{μ}, there corresponds U_{μ} with

$$\alpha < \liminf_{\nu\to\infty} \inf_{u\in U_{\mu}} \sup_{v\in V_{\mu}} F_{\nu}(u,v)$$

For any subsequence $\{\nu_k,\ k=1,\ldots\}$ and any $x_k \xrightarrow{\tau} x$

$$\alpha < \liminf_{k\to\infty} \sup_{v\in V_{\mu}} F_{\nu_k}(x_k,v)$$

because for k sufficiently large $x_k \in U_{\mu}$ and $\liminf_{\nu\to\infty} \le \liminf_{\nu_k\to\infty}$.

This implies the existence of a sequence $\{y_{k\mu},\ k=1,\ldots\}$ such that

$$\alpha < \liminf_{k\to\infty} F_{\nu_k}(x_k,y_{k,\mu}).$$

This being true for any μ, we get

$$\alpha \le \liminf_{\mu\to\infty} \liminf_{k\to\infty} F_{\nu_k}(x_k,y_{k\mu})$$

This and the Diagonalization Lemma A.1 of [7, Appendix] yields a sequence $\{y_k = y_{k,\mu(k)} \in V_k,\ k=1\ldots\}$ such that

$$\alpha \le \liminf_{k\to\infty} F_{\nu_k}(x_k,y_k)$$

and hence

$$\alpha \le \sup_{y_k \xrightarrow{\sigma} y} \liminf_{k\to\infty} F_{\nu_k}(x_k,y_k)$$

Since this holds for any subsequence $\{\nu_k,\ k=1,\ldots\}$ and $x_k \xrightarrow{\tau} x$, $\alpha \le H(x,y)$. This being true for any $\alpha < \text{h/e-li}\, F^{\nu}$, we finally get $\text{h/e-li}\, F^{\nu} \le H$. □

In the metrizable setting it is also possible to characterize the epi/hypo-convergence in terms of the *Moreau-Yosida approximates* [7, Section 5] . Here we review briefly the main results

4.8 DEFINITION. *Let* (X,τ) *and* (Y,σ) *be metrizable, and* d_τ *and* d_σ *metrics compatible with* τ *and* σ *respectively ; and* $F:X\times Y \to \bar{R}$ *a bivariate function. For* $\lambda > 0$ *and* $\mu > 0$ *, the lower Moreau-Yosida approximate (with parameters* λ *and* μ*) is*

$$F^{\downarrow}(\lambda,\mu,x,y) = \sup_{v\in Y} \inf_{u\in X}\left[F(u,v) + \frac{1}{2\lambda}d_\tau^2(u,x) - \frac{1}{2\mu}d_\sigma^2(v,y)\right]$$

and the upper Moreau-Yosida approximate (with parameters λ *and* μ*) is*

$$F^{\uparrow}(\lambda,\mu,x,y) = \inf_{u\in X} \sup_{v\in Y}\left[F(u,v) + \frac{1}{2\lambda}d_\tau^2(u,x) - \frac{1}{2\mu}d_\sigma^2(v,y)\right]$$

4.9 THEOREM. *Suppose* $\{F^\nu, \nu=1,\dots\}$ *is a sequence of extended-real valued bivariate functions defined on the product of the metric spaces* (X,d_τ) *and* (Y,d_σ) *. Suppose there exists* $r > 0$ *and some pair* $(u_o,v_o) \in X\times Y$ *such that* $F^\nu(u_o,v) \le r[d_\sigma^2(v,v_o)+1]$ *and* $F^\nu(u,v) \ge -r[d_\tau^2(u,u_o) + d_\sigma^2(v,v_o)+1]$ *for all* $\nu = 1,\dots$ *. Then*

$$e/h\text{-}ls\, F^\nu(x,y) = \sup_{\lambda>0} \inf_{\mu>0} \limsup_{\nu\to\infty} F_\nu^{\downarrow}(\lambda,\mu,x,y).$$

If there exist r *and* (u_o,v_o) *such that for all* $\nu = 1,\dots$ $F^\nu(u,v_o) \ge -r[d_\tau^2(u,u_o)+1]$ *and* $F^\nu(u,v) \le +r[d_\tau^2(u,u_o)+d_\sigma^2(v,v_o)+1]$ *, then*

$$h/e\text{-}li\, F^\nu(x,y) = \inf_{\mu>0} \sup_{\lambda>0} \liminf_{\nu\to\infty} F_\nu^{\uparrow}(\lambda,\mu,x,y).$$

The Moreau-Yosida approximates [7, Theorem 5.8] are locally equi-Lipschitz, at least when the bivariate functions F^ν can be minorized/majorized as in Theorem 4.9. This is a very useful property ; it allows us to work with well-behaved functions. Moreover, when expressed in terms of the Moreau-Yosida approximates, the epi/hypo-convergence reduces to pointwise limit operations.

5. SEQUENTIAL COMPACTNESS.

The fact that any sequence of bivariate functions, at least in the metrizable case, possesses an epi/hypo-convergent subsequence plays an important role in many applications. One relies on this compactness result to assert the existence of an epi/hypo-limit of a subsequence, then use the specific properties of the elements of the sequence to identify the limit function and finally obtain the epi/hypo-convergence of the whole sequence. In [7], the proof of this compactness theorem is obtain with the help of the Moreau-Yosida approximates and the identities that appear in Theorem 4.9. The proof given here follows the more standard techniques of De Giorgi and Franzoni [11], that such an argument might work was suggested to us by Cavazzuti.

5.1. THEOREM. *Suppose* (X,τ) *and* (Y,σ) *are topological spaces with countable base. Then any sequence of bivariate functions* $\{F^\nu : X\times Y \to \overline{\mathbb{R}},\ \nu = 1,\dots\}$ *contains a subsequence which is* epi_τ/ $hypo_\sigma$*-convergent.*

PROOF. We have to find a subsequence $\{\nu_k,\ k=1,\dots\}$ such that

$$e/h\text{-}ls\ F^{\nu_k} \le h/e\text{-}li\ F^{\nu_k}$$

Let $\{U_\mu | \mu=1,\dots\}$ and $\{V_{\mu'} | \mu'=1,\dots\}$ a countable sequence of open sets in X and Y resp.. From the compactness of $\overline{\mathbb{R}} = [-\infty, +\infty]$ and the classical diagonalization lemma, follows the existence of a subsequence $\{\nu_k | k=1,\dots\}$ such that for every μ and μ'

$$\lim_{k\to 0} \inf_{u\in U_\mu} \sup_{v\in V_{\mu'}} F^{\nu_k}(u,v)$$

and

$$\lim_{k\to 0} \sup_{v\in V_{\mu'}} \inf_{u\in U_\mu} F^{\nu_k}(u,v)$$

exist. It follows that for every μ and μ'

$$\lim_{k\to\infty}\sup \sup_{v\in V_{\mu'}} \inf_{u\in U_\mu} F^{\nu_k}(u,v) \le \lim_{k\to\infty}\inf \inf_{u\in U_\mu} \sup_{v\in V_{\mu'}} F^{\nu_k}(u,v)$$

Hence, for every x and y,

$$\sup_{U_\mu\in \mathcal{N}_\tau(x)} \inf_{V_{\mu'}\in \mathcal{N}_\sigma(y)} \lim_{k\to\infty}\sup \sup_{v\in V_{\mu'}} \inf_{u\in U_\mu} F^{\nu_k}(u,v)$$

$$\le \inf_{V_{\mu'}\in \mathcal{N}_\sigma(y)} \sup_{U_\mu\in \mathcal{N}_\tau(x)} \lim_{k\to\infty}\inf \inf_{u\in U_\mu} \sup_{v\in V_{\mu'}} F^{\nu_k}(u,v)$$

which is the desired inequality. □

6. RELATED NOTIONS TO EPI/HYPO-CONVERGENCE.

Up to now, we have been motivated by the search for a *minimal convergence concept* that allows us to obtain the convergence of saddle points and saddle values, cf. Theorem 2.10. This has led us to a notion of convergence whose limit is not necessarily unique. This is not unexpected, since bivariate functions are not completely determined by their saddle value properties, as already observed by Rockafellar [12] in his work on duality. In the convex-concave setting we formalize this by introducing equivalence classes. The definition of epi/hypo-convergence makes the two variables x and y play a symmetric role, it specializes to epi- or hypo-convergence when the functions are univariate. However, in some applications the F^ν enjoy some continuity properties and it is possible to work with stronger notions of convergence. We exemplify this by giving one such possibility. We proceed as before and start with the definition of two limit functions :

(6.1) $$h_\sigma/e_\tau\text{-ls } F^\nu(x,y) = \inf_{V\in\mathcal{N}_\sigma(y)} \sup_{U\in\mathcal{N}_\tau(x)} \limsup_{\nu\to\infty} \inf_{u\in U} \sup_{v\in V} F^\nu(u,v)$$

and

(6.2) $$e_\tau/h_\sigma\text{-li } F^\nu(x,y) = \sup_{U\in\mathcal{N}_\tau(x)} \inf_{V\in\mathcal{N}_\sigma(y)} \liminf_{\nu\to\infty} \sup_{v\in V} \inf_{u\in U} F^\nu(u,v)$$

We have the following relations :

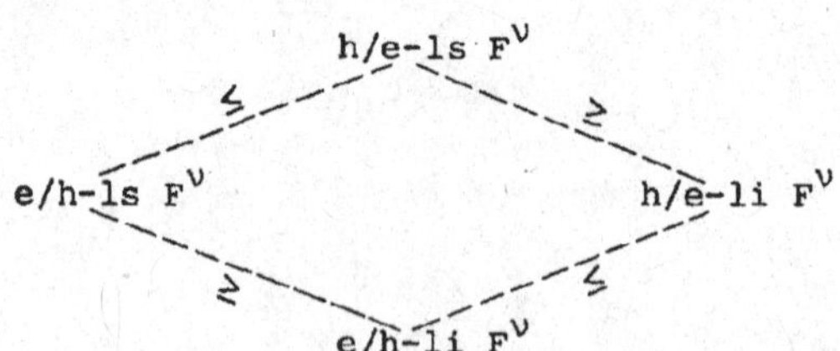

Thus :

$$e/h\text{-}ls\ F^{\nu} = e/h\text{-}li\ F^{\nu}$$

$$h/e\text{-}li\ F^{\nu} = h/e\text{-}ls\ F^{\nu}$$

and, a fortiori,

$$h/e\text{-}ls\ F^{\nu} = e/h\text{-}li\ F^{\nu}$$

imply each epi/hypo-convergence. The convergence induced by the equality $e/h\text{-}ls\ F^{\nu} = e/h\text{-}li\ F^{\nu}$, now with unique limit (τ-l.sc. with respect to x), has been studied by Cavazzuti [8] [9]. The study of the convergence induced by the last equality $h/e\text{-}ls\ F^{\nu} = e/h\text{-}li\ F^{\nu}$, has also been sketched out in [7]. It is possible, for all of these, to develop a theory similar to that for epi/hypo-convergence, but each one of these notions requires a certain regularity for the limit function which, a priori, cannot be guaranteed in many applications.

7. EPI/HYPO-CONVERGENCE OF CONVEX-CONCAVE FUNCTIONS.

This last section is devoted to the continuity properties of the Legendre-Fenchel transform, which establishes a natural correspondence between convex and convex-concave bivariate functions. The argumentation is surprisingly complex, in part this comes from the fact that the functions can take on both the values $+\infty$ and $-\infty$, and that the conjugate operation, or equivalently the Legendre-Fenchel transformation, then looses its local characteristics and it is only the global properties of the operation that are preserved. An elegant study of this phenomena and its implications has been made by Rockafellar [12], [13] and [14] and further analysed by McLinden [15], [16] ; see also Ekeland-Temam [17] and Aubin [18].

Let * denote the *conjugate operation*. For any $F : X \to \overline{R}$ the *conjugate function* is defined by

$$F^*(x^*) = \sup_{x \in X} [\langle x^*, x\rangle - F(x)] .$$

Then one can show [12] that for convex functions

$$F^{**} = \underline{\mathrm{cl}}\, F,$$

where $\underline{\mathrm{cl}}$ F is the *extended closure* of F with

$$\underline{\mathrm{cl}}\, F(x) = \begin{cases} \mathrm{cl}\, F(x) & \text{if } \mathrm{cl}\, F > -\infty \\ -\infty & \text{otherwise} \end{cases}$$

and cl F is the lower semicontinuous closure of F.

Convex-concave bivariate functions can be related to convex bivariate functions through partial conjugation, which means conjugation with respect to one of the variables. We are led to introduce equivalence classes. For the sake of the uninformed reader we review quickly the motivation and the main features of Rockafellar's scheme [13].

Let K_o be convex-concave continuous function on $[-1,1] \times [-1,1]$. We associate to K_o the two functions :

$$K_1(x,y) = \begin{cases} +\infty \text{ if } |x| > 1 \\ K_o(x,y) \text{ on } [-1,1]\times[-1,1] \\ -\infty \text{ if } |x|\leq 1 \text{ and } |y| > 1 \end{cases}$$

and

$$K_2(x,y) = \begin{cases} +\infty \text{ if } |x| > 1 \text{ and } |y| \geq 1 \\ K_o(x,y) \text{ on } [-1,1]\times[-1,1] \\ -\infty \text{ if } |y| > 1. \end{cases}$$

Then both K_1 and K_2 have the same saddle points (and values) as K_o, although they differ on substantial portions of the plane. However, not only do these two functions have the same saddle points but so do all linear perturbations of these two functions. So from a variational viewpoint these two functions appear to be undistinguishable. It is thus natural when studying limits of a variational character that we need to deal with equivalence classes whose members have similar saddle point properties.

Let $K : X\times Y \to \bar{R}$ be a convex-concave function. We associate to K its convex and concave parents defined by

$$F(x,y^*) = \sup_{y \in Y} [K(x,y) + \langle y^*,y\rangle]$$

and

$$G(x^*,y) = \inf_{x \in X} [K(x,y) - \langle x^*,x\rangle].$$

Thus we have the following relations between these functions

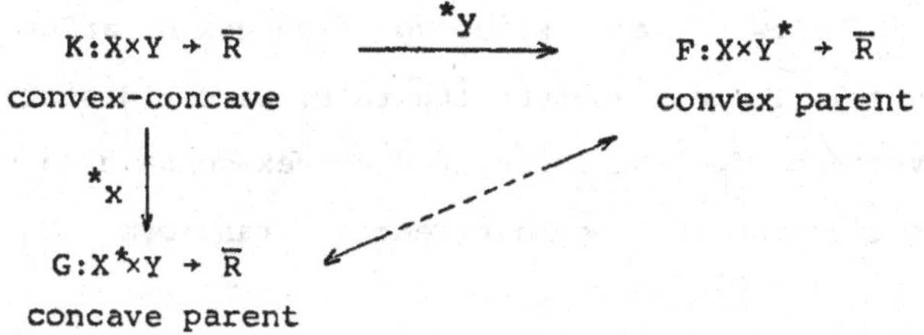

In the example above K_1 and K_2 have the same parents, they cannot be distinguished as coming from different bivariate convex or bivariate concave functions.

Given any pair of convex-concave bivariate functions K_1 and K_2, we say that they are *equivalent* if they have the same parents. A bivariate function K is said to be *closed* if its parents are conjugates of each other, i.e. if the above diagram can be closed through the classical Legendre-Fenchel transform :

$$-G(x^*,y) = \sup_{\substack{x \in X \\ y^* \in Y^*}} [\langle x,x^*\rangle + \langle y,y^*\rangle - F(x,y^*)]$$

For closed convex-concave functions K, the associated equivalence class is an interval, denoted by $[\underline{K},\overline{K}]$ with

$$\underline{K} = \underline{cl}_x K = \sup_{x^* \in X} [G(x^*,y) + \langle x,x^*\rangle] ,$$

and

$$\overline{K} = \overline{cl}_y K = \inf_{y^* \in Y} [F(x,y^*) - \langle y,y^*\rangle] ,$$

where $\underline{cl}_x$ denotes the extended lower closure with respect to x and $\overline{cl}_y K (= - \underline{cl}_y(-K))$ is the extended upper closure with respect to y. A convex function $F:X\times Y^* \to \overline{R}$ is *closed* if $\underline{cl}_{(x,y^*)}F = F$.

7.1 THEOREM. [13] .*The map* $K \overset{*y}{\longmapsto} F$ *establishes a one-to-one correspondence between closed convex-concave (equivalence) classes and closed convex functions.*

This correspondence $*y$ has continuity properties that are made explicit here below. Given a sequence of convex bivariate functions $\{F^\nu, \nu = 1,\ldots\}$ epi_τ-converging to F, we could study the induced convergence for the associated convex-concave (bivariate) functions (through the Legendre-Fenchel transform $^*y$).

In the reflexive Banach case, it would be natural to consider epi_τ-convergence to be epi-convergence induced by the weak and the strong topologies on X×Y. To illustrate these type of results, we consider the situation when the $\{F^\nu, \nu = 1,\ldots\}$ epi-converge to F with respect to the Mosco-topology.

We start with a quick review of Mosco-epi-convergence or for short, Mosco-convergence. Suppose X and Y are reflexive Banach spaces whose weak and strong topologies are denoted by w_X, w_Y, s_X, s_Y respectively. A sequence of functions, which for reasons of exposition we take here as bivariate,

$$\{F^\nu: X\times Y \to \bar{R}\ ,\ \nu = 1,\ldots\}$$

is said to *Mosco-converge* to the (bivariate) function

$$F : X\times Y \to \bar{R}$$

if

$$e_s\text{-}\mathrm{ls}\ F^\nu \le F \le e_w\text{-}\mathrm{li}\ F^\nu$$

where

$$e_s\text{-}\mathrm{ls}\ F^\nu(x,y) = e_{s_X\times s_Y}\text{-}\mathrm{ls}\ F^\nu(x,y)$$

$$= \sup_{(U,V)\in \mathcal{N}_{s_X}(x)\times\mathcal{N}_{s_X}(y)} \limsup_{\nu\to\infty} \inf_{u\in U, v\in V} F^\nu(u,v)$$

and

$$e_w\text{-}\mathrm{li}\ F^\nu(x,y) = e_{w_X\times w_Y}\text{-}\mathrm{li}\ F^\nu(x,y)$$

$$= \sup_{(U,V)\in \mathcal{N}_{w_X}(x)\times\mathcal{N}_{w_Y}(y)} \liminf_{\nu\to\infty} \inf_{u\in U, v\in V} F^\nu(u,v)\,.$$

Because of the natural relations between epi-limits, this means that

$$F = e_w\text{-}\mathrm{li}\ F^\nu = e_s\text{-}\mathrm{li}\ F^\nu = e_w\text{-}\mathrm{ls}\ F^\nu = e_s\text{-}\mathrm{ls}\ F^\nu.$$

This type of convergence has been introduced by Mosco [19] and studied extensively because of the role it plays in many applications, cf. for example [19], [20] and [10]. The basic result for convex functions first proved by Wijsman [21] in the finite dimensional case and extended by Mosco and Joly [22] to the Banach reflexive case is the bicontinuity of the Legendre-Fenchel transform with respect to the Mosco-topology. In our setting, we can express this result through the identity

(7.2) $$(e_w\text{-li } F^\nu)^* = e_s\text{-ls } F^{\nu *}$$

where * denotes conjugation with respect to both variables. This is a special case of the more general relation that we need between convex bivariate functions and classes of convex-concave bivariate functions. We only sketch the proof whose details appear in [23].

7.3. THEOREM. X *and* Y *are reflexive Banach spaces, and* $\{F; F^\nu : X \times Y \to \overline{R},\ \nu = 1,\ldots\}$ *is a collection of closed proper convex functions. Let* $[\overline{K}, \underline{K}]$ *and* $\{[\overline{K}^\nu, \underline{K}^\nu],\ \nu = 1,\ldots\}$ *be the corresponding classes of bivariate convex-concave functions. Then, the following statements are equivalent :*

(7.4) *the* F^ν *Mosco-converge to* F

and

(7.5) *for all* K *in* $[\overline{K}, \underline{K}] \neq \emptyset$ *, we have*

$$\underline{cl}_x(e_s/h_w\text{-ls } K^\nu) \leq K \leq \overline{cl}^y(h_s/e_w\text{-li } K^\nu).$$

PROOF (Sketch). The key step consists in extending the result about the bicontinuity of the Legendre-Fenchel transform for convex functions to this setting, i.e. for bivariate convex functions and partial conjugation. In particular one shows that if the F and F^ν are a collection of bivariate closed proper convex functions, and $F = e_{w(X\times Y^*)}\text{-lim } F^\nu$ then

(7.6) $$\bar{K} = \overline{cl}_y(h_s/e_w\text{-li } \bar{K}^\nu)$$

and

(7.7) $$\underline{K} = \underline{cl}_x(h_s/e_w\text{-li } \bar{K}^\nu)$$

The proof of the inequality $\bar{K} \geq \overline{cl}_y(h_s/e_w\text{-li } \bar{K}^\nu)$ follows directly from the definitions of epi- and epi/hypo-convergence and the Legendre-Fenchel transform. The converse inequality is much more difficult to obtain. One start with deriving

(7.8) $$\underline{K} \leq h_s/e_w\text{-li } \bar{K}^\nu,$$

which is first obtained under the additional condition that the F and F^ν , $\nu = 1,\ldots$ are equi-coercive. To bring the general problem in this more restrictive framework, we rely on Moreau-Yosida approximates. The F^ν are replaced by

$$F^{\nu,\lambda}(x,y^*) = F^\nu(x,y^*) + \frac{\lambda}{2}|y^*|^2$$

and F by

$$F^\lambda(x,y^*) = F(x,y^*) + \frac{\lambda}{2}|y^*|^2 .$$

For each $\lambda > 0$, we then have

$$\bar{K}^\lambda(x,y) \leq h_s/e_w\text{-li } \bar{K}^{\nu,\lambda}$$

We then use the monotonocity in λ, and a diagonalization lemma [7, Lemma A.1] to conclude that

$$\bar{K}(x,y) \leq \liminf_{\nu\to\infty} \bar{K}^{\nu,\lambda(\nu)}(x_\nu,y)$$

for all $x_\nu \xrightarrow[w]{} x$ and $\lambda(\nu)$ some subsequence converging to 0 as $\nu \to \infty$. Using the properties of conjugation this allows us to obtain the existence of $\{y_\nu,\ \nu = 1,\ldots\}$ such that

$$\bar{K}(x,y) \leq \liminf \left[\bar{K}^\nu(x_\nu,y_\nu) - \frac{1}{2\lambda(\nu)}|y - y_\nu|^2\right].$$

There remains only to show that the sequence $\{\overline{K}^\nu(x_\nu,y_\nu), \nu = 1,\dots\}$ is bounded above. This is equivalent to the assertion that for every weakly convergent sequence $\{x_\nu, \nu = 1,\dots\}$ there exists a bounded sequence $\{y^*_\nu, \nu = 1,\dots\}$ such that $\sup_\nu F^\nu(x_\nu,y^*_\nu) < +\infty$. This would impose a very strong restriction on the sequence $\{F^\nu, \nu = 1,\dots\}$. To avoid imposing any such condition, rather than working with the F^ν and F, we work with the F^ν_μ and F_μ that are the Moreau-Yosida approximates with respect to x of the F^ν and F respectively, i.e.

$$F^\nu_\mu(x,y^*) = \inf_{u\in X}\left[F^\nu(u,y^*) + \frac{1}{2\mu}|u - x|^2\right]$$

and

$$F_\mu(x,y^*) = \inf_{u\in X}\left[F(u,y^*) + \frac{1}{2\mu}|u - x|^2\right]$$

The desired inequality is then obtained for the $\overline{K}_\mu$ and $\overline{K}^\nu_\mu$, the (partial) Legendre-Fenchel transforms of the F_μ and F^ν_μ. It is then shown that

$$\sup_{\mu>0} \overline{K}_\mu = \underline{cl}_x\, \overline{cl}_y\, K = \underline{K}$$

which allows us to obtain (7.8) without any coercivity restrictions on the function F^ν and F.

We now return to the core of the proof of the Theorem. To say that the $\{F^\nu, \nu = 1,\dots\}$ Mosco-converge to F means that

$$e_s\text{-ls } F^\nu \le F \le e_w\text{-li } F^\nu$$

The second inequality, through (7.6) yields

(7.9) $$\overline{K} \le \overline{cl}_y(h_s/e_w\text{-li } \overline{K}^\nu).$$

The first inequality yields

$$\underline{K} \le \overline{cl}_x(e_w/h_s\text{-ls } \underline{K}^\nu)$$

through (7.7) and using this time the following arguments : the inequality $e_s\text{-ls}\, F^\nu \le F$ implies

$$e_w\text{-li}(-G^\nu) \ge -G$$

with G the concave conjugate of F as identified in the diagram above ; and then G is used in the construction of $\underline{K}$ (and G^ν for $\underline{K}^\nu$). The reasoning is totally symmetric.

There remains to show that (7.5) implies (7.4). By exploiting duality and the fact that we are working with closed convex-concave functions, one can prove that it really will suffice to show that for all $\{(x_\nu, y_\nu^*),\ \nu = 1,\ldots\}$ that weakly converge to (x, y^*) we have

(7.10) $$F(x,y^*) \le \liminf_{\nu\to\infty} F^\nu(x_\nu, y_\nu^*).$$

Since $K \le \overline{\text{cl}}^Y(h_s/e_w\text{-li}\,\overline{K}^\nu)$, we have that for every (x,y^*)

$$\begin{aligned} F(x,y^*) &= \sup_{y\in Y}\Big[\langle y^*,y\rangle + K(x,y)\Big] \\ &\le \sup_{y\in Y}\Big[\langle y^*,y\rangle + \overline{\text{cl}}^Y(h_s/e_w\text{-li}\,\overline{K}^\nu)\Big] \\ &\le \sup_{y\in Y}\Big[\langle y^*,y\rangle + h_s/e_w\text{-li}\,\overline{K}^\nu\Big]. \end{aligned}$$

Thus to prove (7.10) it suffices to show that for any sequence $(x_\nu, y_\nu^*) \xrightarrow[w]{} (x,y^*)$ and any $y \in Y$

$$\langle y^*,y\rangle + h_s/e_w\text{-li}\,\overline{K}^\nu(x,y) \le \liminf_{\nu\to\infty} F^\nu(x_\nu, y_\nu^*).$$

Using the definition of epi/hypo-convergence, in particular of $h_s/e_w\text{-li}\,\overline{K}^\nu$, we see that to each $x_\nu \xrightarrow[w]{} x$ we can associate a strongly convergent sequence $y_\nu \xrightarrow[s]{} y$ such that

$$\langle y^*,y\rangle + h_s/e_w\text{-li}\,\overline{K}^\nu(x,y) \le \langle y^*,y\rangle + \liminf_{\nu\to\infty} \overline{K}^\nu(x_\nu, y_\nu)$$

and thus we only need to show that

$$\langle y^*,y\rangle + \liminf_{\nu\to\infty} \overline{K}^\nu(x_\nu,y_\nu) \le \liminf_{\nu\to\infty} F^\nu(x_\nu, y_\nu^*).$$

But this follows directly from the relation

$$F^{\nu}(x_\nu, y_\nu^*) = \sup_{y \in Y} \{\langle y_\nu^*, y\rangle + \bar{K}^{\nu}(x_\nu, y)\}$$

$$\geq \langle y_\nu^*, y_\nu\rangle + \bar{K}^{\nu}(x_\nu, y_\nu). \quad \square$$

APPENDIX.

We give here the proof that we take from [24] , of a diagonalization result used in the proof of Theorem 4.1.

A.1. LEMMA (Diagonalization). Suppose $\{a_{\nu,\mu} | \nu = 1,\dots\}$ is a doubly indexed family of extended-reals numbers. Then there exists a map $\nu \longrightarrow \mu(\nu)$ increasing such that

$$\text{(A.2)} \qquad \limsup_{\mu\to\infty} \liminf_{\nu\to\infty} a_{\nu,\mu} \geq \liminf_{\nu\to\infty} a_{\nu,\mu(\nu)}$$

PROOF. Let us denote $a_\mu = \liminf_{\nu\to+\infty} a_{\nu,\mu}$ and $a = \limsup_{\mu\to+\infty} a_\mu$

If $a = +\infty$ there is nothing to prove. So, let us assume that $a < +\infty$. By definition of a, there exists an increasing sequence $\{\mu_p ; p = 1,2\dots\}$, $\mu_p \xrightarrow[p\to+\infty]{} +\infty$ such that

$$\sup(-p, a + 2^{-p}) \geq a_\mu \quad \text{for all } \mu \geq \mu_p$$

By definition of a_μ, there exists an increasing sequence $\{\nu_p ; p = 1,2\dots\}$, $\nu_p \xrightarrow[p\to+\infty]{} +\infty$ such that

$$\sup(-p, a_{\mu_p} + 2^{-p}) \geq a_{\nu_p,\mu_p} \quad \text{for all } p \in \mathbb{N}.$$

Let us define $\mu(\nu) = \mu_p$ if $\nu_p \leq \nu \leq \nu_{p+1}$

Then, $\liminf_{\nu\to+\infty} a_{\nu,\mu(\nu)} \leq \liminf_{p\to+\infty} a_{\nu_p,\mu(\nu_p)} = \liminf_{p\to+\infty} a_{\nu_p,\mu_p}$, as follows from the definition of $\mu(\nu_p) = \mu_p$.

From the two above inequalities, we derive

$$\liminf_{\nu\to+\infty} a_{\nu,\mu(\nu)} \leq \liminf_{p\to+\infty} \left[\sup(-p, \sup(-p, a+2^{-p}) + 2^{-p})\right]$$

$$\leq a \quad . \square$$

REFERENCES.

[1] Glowinski, R., Lions, J.L., and Trémolières, R., Analyse Numérique des Inéquations Variationnelles, Tome 1, Dunod.

[2] De Giorgi, E., and Spagnolo, S., Sulla convergenza degli integrali dell'energia per operatori ellittici del 2e ordine, Boll. Un. Mat. Ital., (4) 8 (1973, 391-411.

[3] Tartar, L., Cours Peccot, Mars 1979, Collège de France; and Murat, F., H-convergence, Seminaire Alger, 1977-78.

[4] Fiacco, A., and Hutzler, W., Optimal value differential stability results for general inequality constrained differentiable mathematical programs, in Mathematical Programming with Data Perturbations, I, ed. A. Fiacco, Marcel Dekker, New-York, 1982, 29-44.

[5] Zolezzi, T., On the stability analysis in mathematical programming, Manuscrip, IMA-Genova, 1981.

[6] De Giorgi, E., Convergence problems for functionals and operators, in Recent Methods in Nonlinear Analysis, Pitagora Editrice, Bologna, 1979, pp. 131-138.

[7] Attouch, H., and Wets, R., A convergence theory for saddle functions, Trans. Amer. Math. Soc. (1983).

[8] Cavazzuti, E., Γ-convergence multiple, convergenza di punti di sella e di max-min, Boll. Un. Mat. Ital., to appear.

[9] Cavazzuti,E., Alcune caraterizzazioni della Γ-convergenza multipla, Manuscrip, Università di Modena, 1981.

[10] Sonntag, Y., Convergence au sens de U. Mosco : théorie et applications à l'approximation des solutions d'inéquations, Thèse, Univ. Provence, 1982.

[11] De Giorgi, E., and Franzoni, T., Su un tipo di convergenza variazionale, Atti Acc. Naz. Lincei (8), 58 (1975), 842-850. See also : Fascicolo delle demostrazioni, Scuola Normale Superiore di Pisa, May 1976.

[12] Rockafellar, R.T., A general correspondence between dual minimax problems and convex programs, Pacific J. Math., 25 (1968), 597-611.

[13] Rockafellar, R.T., Monotone operators associated with saddle functions and minimax problems, in Nonlinear Functional Analysis, ed. F. Browder, Amer. Math Soc. Providence, 1970.

[14] Rockafellar, R.T., Convex Analysis, Princeton University Press, Princeton, 1970.

[15] McLinden, L., A minimax theorem, Math. Operations Res., (to appear).

[16] McLinden, L., Dual operations on saddle functions, Trans. Amer. Math. Soc., 179 (1973), 363-381.

[17] Ekeland, I., and Temam, R., Analyse Convexe et Problèmes Variationnels, Dunod, Paris, 1974.

[18] Aubin, J.P., Mathematical methods of game and economic theory, North-Holland, Amsterdam, 1979.

[19] Mosco, U., Convergence of convex sets and of solutions of variational inequalities, Advances Math., 3 (1969), 273-299.

[20] Attouch, H., Familles d'opérateurs maximaux monotones et mesurabilité, Annali di Matematica pura ed applicata (IV), CXX (1979), 35-111.

[21] Wijsman, R., Convergence of sequences of convex sets, cones and functions, II, Trans. Amer. Math. Soc., 123 (1966), 32-45.

[22] Joly, J.L., Une famille de topologies et de convergences sur l'ensemble des fonctionelles convexes, Thèse Grenoble, France, 1970.

[23] Attouch, H., and Wets, R., Convergence of convex-concave saddle functions, to appear.

[24] Attouch, H., Variational Convergence for Functions and Operators, Research Notes in Mathematics, Pittman, London, to appear.

OPTIMAL FEEDBACK CONTROLS FOR SEMILINEAR PARABOLIC EQUATIONS

By

Viorel Barbu

University of Iaşi, Romania

1. Introduction

We are here concerned with existence of an optimal feedback control for the system

$$
\begin{aligned}
&y_t+Ay+\beta(y) = Bu && \text{in } Q=\Omega\times\,]o,T[\\
(1.1)\quad &y(x,o) = y_o(x) && \text{a.e. } x\in\Omega\\
&y(x,t) = o && \text{for } (x,t)\in\Sigma=\Gamma\times\,]o,T[
\end{aligned}
$$

with the cost functional

$$(1.2)\quad \int_o^T (\varphi(y(t))+h(u(t)))dt+\psi(y(T))$$

where

1° Ω is an open and bounded subset of the Euclidean space R^N with a sufficiently smooth boundary Γ, and A is a second order elliptic differential operator on Ω of the form

$$(1.3)\quad Ay=-\sum_{i,j=1}^{N}(a_{ij}(x)y_{x_i})_{x_j}+a_o(x)y\ ,$$

where $a_{ij}\in C^1(\bar\Omega)$, $a\in L^\infty(\Omega)$, $a\geq o$ a.e. on Ω, $a_{ij}=a_{ji}$ for all $i,j=1,\dots N$ and for some $\omega>o$,

$$(1.4)\quad \sum_{i,j=1}^{N} a_{ij}(x)\zeta_i\zeta_j\geq\omega|\zeta|^2 \quad \text{a.e. } x\in\Omega\ ,\quad \zeta\in R^n.$$

2° B is a linear continuous operator from a real Hilbert space U to $L^2(\Omega)$. The norm of U will be denoted by $\|\cdot\|$ and the inner product by $\langle \cdot,\cdot \rangle$.

3° β is a continuous increasing function on R such that $0 = \beta(0)$. Let $j: R \longrightarrow R$ be a continuous convex function such that $\partial j = \beta$ (∂j is the subdifferential of j.)

4° The functions $\Phi : L^2(\Omega) \longrightarrow R$ and $\Psi : L^2(\Omega) \longrightarrow R$ are locally Lipschitz and positive.

5° The function $h: U \longrightarrow \bar{R} =]-\infty, +\infty]$ is convex, lower semicontinuous, $\not\equiv +\infty$ and satisfies the growth condition ($C_1 > 0$)

$$h(u) \geq C_1 \|u\|^2 + C_2 \quad \text{for all } u \in U.$$

From now on we shall denote by H the space $L^2(\Omega)$ with the usual scalar product $(\cdot,\cdot)$ and norm $|\cdot|$. The norm in the space R^n will be denoted by $|\cdot|_n$. By a locally Lipschitz function we mean a function which is Lipschitz on bounded sets.

Given a locally Lipschitz function $\varphi : H \longrightarrow R$ we shall denote by $\varphi^0 : H \times H \longrightarrow R$ the function

$$(1.5) \qquad \varphi^0(y,v) = \limsup_{\substack{\lambda \to 0 \\ z \to y}} (\varphi(z+\lambda v) - \varphi(z))\lambda^{-1}$$

and by $\partial \varphi : H \longrightarrow 2^H$ the generalized gradient of φ ([7],[8],[13])

$$\partial \varphi(y) = \{ w \in H; (w,v) \leq \varphi^0(y,v) \quad \text{for all } v \in H \}.$$

If φ is convex then $\partial \varphi$ is just the subdifferential of φ . If φ admits a continuous Gâteaux derivative $\nabla \varphi$ then $\partial \varphi = \nabla \varphi$.

In the sequel we shall denote by $H^{2,1}(Q_t)$, $Q_t = \Omega \times]t,T[$ the usual Sobolev space $\{ y \in L^2(t,T;H_0^1(\Omega) \cap H^2(\Omega)); y_s \in L^2(t,T;H) \}$

where y_s is the derivative of y as a function of s from $[t,T]$ to H. For $t=0$ we shall simply write Q instead of Q_0. We set $Q^0=\Omega\times R^+$ and denote by $H^{2,1}_{loc}(Q^0)$ the space $\{y\in L^2_{loc}(R^+;H^1_0(\Omega)\cap H^2(\Omega));$ $y'\in L^2_{loc}(R^+;H)\}$ where $R^+=[0,+\infty[$ and y' is the strong derivative of y.

Given a Banach space X and $[a,b]$ a compact interval, we shall denote by $C([a,b];X)$ the space of all continuous X-valued functions on $[a,b]$ and by $BV([a,b];X)$ the space of X-valued functions of bounded variation on $[a,b]$. By $W^{1,2}(a,b;X)$ we shall denote the space $\{y\in L^2(a,b;X);y'\in L^2(a,b;X)\}$ and by $AC([a,b];X)$ the space of all absolutely continuous functions $y:[a,b]\longrightarrow X$.

By $AC_{loc}(R^+;X)$ we shall denote the space of functions $y:R^+\longrightarrow X$ which are absolutely continuous on every compact of R^+. Let $F:H\longrightarrow H$ be defined by

(1.6) $\quad Fy = Ay + \beta(y) \qquad \text{for} \quad y\in D(F)$

(1.7) $\quad D(F)=\{y\in H^1_0(\Omega)\cap H^2(\Omega);\ \beta(y)\in L^2(\Omega)\}$

It is useful to notice that

(1.8) $\quad (Fy,y)\geq \omega|y|^2 \qquad \text{for all} \quad y\in D(F).$

In terms of F we may write (1.1) as

$$(1.1)'\qquad \begin{cases} y'+Fy\ni Bu & \text{a.e. } t\in\,]0,T[\\ y(0)=y_0\,. & \end{cases}$$

We recall (see for instance [1] p.202) that if $y_0\in H$ and $u\in L^2(t,T;U)$ then Eq.(1.1) ((1.1)') with initial condition $y(t,.)=y_0$ has a unique solution $y(s)=y(s,t,y_0,u)$ on Q_t which satisfies

(1.9) $(s-t)^{1/2}y \in L^2(t,T;H_0^1(\Omega) \cap H^2(\Omega));(s-t)^{1/2}y_s \in L^2(t,T;H)$.

If $y_0 \in H_0^1(\Omega)$ and $j(y_0) \in L^1(\Omega)$ then $y \in H^{2,1}(Q_t)$.
We shall denote by $\varphi : [0,T] \times H \longrightarrow R$ the <u>value function</u> associated to problem (1.1), (1.2), i.e.,

$$(1.10) \quad \varphi(t,y_0) = \inf\{ \int_t^T (\Phi(y(s,t,y_0,u))+h(u(s)))ds + \\ + \Psi(y(T,t,y_0,u));u \in L^2(t,T;U) \}.$$

It follows by (1.6) that for each $t \in [0,T]$ and $y_0 \in H$ the map $u \longrightarrow (s-t)^{1/2}y(s,t,y_0,u)$ is bounded from $L^2(t,T;U)$ to $H^{1,2}(Q_t)$ and therefore by the Arzela-Ascoli theorem it is compact from $L^2(t,T;U)$ to $C([t,T];H)$. This implies by a standard device that for each $(t,y_0) \in [0,T] \times H$ the infimum defining $\varphi(t,y_0)$ is attained.

The contents of the paper are outlined below.

In section 2 we shall derive necessary conditions of optimality for problem (1.1), (1.2) (Euler-Lagrange equations) in terms of generalized gradients of Φ, Ψ and β. In section 3 it will be proved that

$$u(t) \in -\partial \varphi(t,y(t)) \qquad 0 \leq t \leq T$$

is an <u>optimal feedback control</u> for problem (1.1), (1.2). Furthermore, it is shown that the value function φ is the solution to a certain Hamilton-Jacobi equation. In section 4 it is studied problem (1.1), (1.2) in the case $T = +\infty$ and in section 5 are given some applications to the time optimal problem associated with system (1.1).

2. Necessary conditions for optimality

We shall study here the optimal control problem with state system (1.1) and cost (1.2), i.e.,

$$(2.1)\quad \inf\{\int_0^T (\Phi(y(s,o,y_o,u))+h(u(s)))ds+\Psi(y(T,o,y_o,u))\ ;\ u\in L^2(o,T;U)\}$$

where

$$(2.2)\quad y_o\in H_o^1(\Omega),\ j(y_o)\in L^1(\Omega)$$

and the functions $\Phi: H\longrightarrow R$, $\Psi: H\longrightarrow R$, $h: U\longrightarrow R$ and A, β satisfy conditions $1^o\sim 5^o$ of section 1.

Let $(y^*,u^*)\in H^{2,1}(Q)\times L^2(o,T;U)$ be an arbitrary optimal pair of problem (2.1).

For any $\varepsilon>o$ consider the control problem: minimize

$$(2.3)\quad \int_0^T (\Phi^\varepsilon(y(t))+h_\varepsilon(u(t))+\frac{1}{2}\|u(t)-u^*(t)\|^2)dt+\Psi^\varepsilon(y(T))$$

over all $u\in L^2(o,T;U)$ and $y\in H^{2,1}(Q)$ subject to

$$(2.4)\quad \begin{aligned} &y_t+Ay+\beta^\varepsilon(y)=Bu && \text{in } Q\\ &y(o)=y_o && \text{in } \Omega\\ &y=o && \text{in } \Sigma \end{aligned}$$

where

$$(2.5)\quad h_\varepsilon(u)=\inf\{(2\varepsilon)^{-1}\|u-v\|^2+h(v);v\in U\}$$

$$(2.6)\quad \beta^\varepsilon(r)=\int_{-\infty}^{\infty}\beta_\varepsilon(r-\varepsilon\tau)\rho(\tau)d\tau\ .$$

Here $\beta_\varepsilon=\varepsilon^{-1}(1-(1+\varepsilon\beta)^{-1})$ and ρ is a C_o^∞- "mollifier" on R. The functions Φ^ε and Ψ^ε are defined as follows. Let $\{e_i\}$ be an orthonormal basis in H and let X_n be the linear space generated by $\{e_i\}_{i=1}^{n}$. For $n = [\varepsilon^{-1}]$ we define

$$(2.7)\qquad \Phi^{\varepsilon}(y) = \int_{R^n} \Phi(P_n y - \varepsilon A_n \tau)\rho_n(\tau)d\tau, \quad y \in H$$

and

$$(2.8)\qquad \Psi^{\varepsilon}(y) = \int_{R^n} \Psi(P_n y - \varepsilon A_n \tau)\rho_n(\tau)d\tau, \quad y \in H$$

where $P_n: H \longrightarrow X_n$ is the projection operator on X_n, ρ_n is a C_o^{∞}- "mollifier" in R^n and $A_n: R^n \longrightarrow X_n$ is the operator

$$(2.9)\qquad A_n(\tau) = \sum_{i=1}^{n} \tau_i e_i \; ; \quad \tau = (\tau_1, \dots \tau_n).$$

Clearly the functions Φ^{ε}, Ψ^{ε}, h_{ε} are Lipschitz and Fréchet differentiable on H.

Let $(y_{\varepsilon}, u_{\varepsilon})$ be an optimal pair for problem (2.3).

LEMMA 1. *For* $\varepsilon \longrightarrow o$ *one has*

(2.10) $u_{\varepsilon} \longrightarrow u^*$ strongly in $L^2(o,T;U)$

(2.11) $y_{\varepsilon} \longrightarrow y^*$ strongly in $C([o,T];H)$ and weakly in $H^{2,1}(Q)$.

Proof. We have

$$(2.12)\qquad \int_o^T (\Phi^{\varepsilon}(y_{\varepsilon}) + h_{\varepsilon}(u_{\varepsilon}) + \tfrac{1}{2}\|u_{\varepsilon} - u^*\|^2)dt + \Psi^{\varepsilon}(y_{\varepsilon}(T)) \le \int_o^T (\Phi^{\varepsilon}(z_{\varepsilon}) + h(u^*))dt + \Psi^{\varepsilon}(z_{\varepsilon}(T))$$

where $z_{\varepsilon} \in H^{2,1}(Q)$ is the solution to

$$(z_{\varepsilon})_t + Az_{\varepsilon} + \beta^{\varepsilon}(z_{\varepsilon}) = Bu^* \quad \text{in } Q$$
$$z_{\varepsilon}(o) = y_o \quad \text{in } \Omega$$

Recalling that $z_{\varepsilon} \longrightarrow y^*$ in $C([o,T];H)$ (see 3) we have

$$(2.13)\quad \lim_{\varepsilon\to 0} \int_0^T dt \int_{R^n} \Phi(P_n z_\varepsilon - \varepsilon A_n \tau)\, \rho_n(\tau)d\tau = \int_0^T \Phi(y^*)dt$$

and

$$(2.14)\quad \lim_{\varepsilon\to 0} \Psi^\varepsilon(z_\varepsilon(T)) = \Psi(y^*(T)).$$

On the other hand, it follows by (2.12) that u_ε is bounded in $L^2(o,T;U)$. Hence on some subsequence again denoted ε we

$u_\varepsilon \longrightarrow u_1$ weakly in $L^2(o,T;U)$

$y_\varepsilon \longrightarrow y_1$ weakly in $H^{2,1}(Q)$ and strongly in $C([o,T];H)$

and therefore

$$(2.15)\quad \lim_{\varepsilon\to 0} \int_0^T \Phi^\varepsilon(y_\varepsilon)dt = \int_0^T \Phi(y_1)dt$$

$$(2.16)\quad \lim_{\varepsilon\to 0} \Psi^\varepsilon(y_\varepsilon(T)) = \Psi(y_1(T)).$$

Finally, by the Fatou lemma,

$$(2.17)\quad \liminf_{\varepsilon\to 0} \int_0^T h_\varepsilon(u_\varepsilon)dt \geqslant \int_0^T h(u_1)dt.$$

Along with (2.12), (2.13), (2.14) formulas (2.15), (2.16), (2.17) imply (2.10). As regards (2.11) it follows by (2.10).

Using the fact that the functions Φ^ε, Ψ^ε, h_ε and β^ε are differentiable it follows by a standard device that there exists $p_\varepsilon \in H^{2,1}(Q)$ which satisfies the system

$$(2.18)\quad (p_\varepsilon)_t - Ap_\varepsilon - p_\varepsilon \nabla\beta^\varepsilon(y_\varepsilon) = \nabla\Phi^\varepsilon(y_\varepsilon) \quad \text{in } Q$$

$$(2.19)\quad p_\varepsilon = o \quad \text{in } \Sigma$$

$$(2.20)\quad p_\varepsilon(T) + \nabla\Psi^\varepsilon(y_\varepsilon(T)) = o \quad \text{a.e. on } \Omega$$

$$(2.21)\quad B^* p_\varepsilon(t) = \nabla h_\varepsilon(u_\varepsilon(t)) + u_\varepsilon(t) - u^*(t) \quad \text{a.e. } t \in]o,T[$$

Since $\nabla \Phi^{\varepsilon}(y_{\varepsilon})$ is bounded in $C([o,T];H)$ and $\nabla \beta^{\varepsilon} \geq o$ we get by (2.18) and (2.20) the following estimates

(2.22) $\quad |p_{\varepsilon}(t)| + \|p_{\varepsilon}\|_{L^2(o,T;H_o^1(\Omega))} \leq C \quad$ for $\quad t \in [o,T]$

(2.23) $\quad \|\nabla \beta^{\varepsilon}(y_{\varepsilon})p_{\varepsilon}\|_{L^1(Q)} \leq C.$

In particular $(p_{\varepsilon})_t$ is bounded in the space $L^1(o,T;H^{-s}(\Omega))$ where $s > N/2$. Since the injection of $L^2(\Omega)$ into $H^{-s}(\Omega)$ is compact, by a theorem of Helly there is a subsequence of p_{ε} which converges pointwise to a limit $p \in BV([o,T];H^{-s}(\Omega))$ in the strong topology of $H^{-s}(\Omega)$. In other words, we may assume that

(2.24) $\quad p_{\varepsilon}(t) \longrightarrow p(t)$ strongly in $H^{-s}(\Omega)$ for $t \in [o,T]$.

On the other, for every $\mu > o$ there is $C(\mu) > o$ such that (see [11] Chap.I, Lemma 5.1)

$$|p_{\varepsilon}(t)-p(t)| \leq \mu\|p_{\varepsilon}(t)-p(t)\|_{H_o^1(\Omega)} +$$

$$+ C(\mu)\|p_{\varepsilon}(t)-p(t)\|_{H^{-s}(\Omega)}, \quad t \in [o,T].$$

Along with (2.24) the latter implies that

(2.25) $\quad p_{\varepsilon} \longrightarrow p$ strongly in $L^2(Q)$ and weakly in $L^2(o,T;H_o^1(\Omega))$

and

(2.26) $\quad p_{\varepsilon}(t) \longrightarrow p(t)$ weakly in H for all $t \in [o,T]$.

By (2.23) we see that on some subsequence we have

(2.27) $\quad \mu_p = \lim_{\varepsilon \to o} \nabla\beta^{\varepsilon}(y_{\varepsilon})p_{\varepsilon}$ weak star in $M(Q)$

where $M(Q)$ is the space of all bounded Radon measures on Q. Summarising at this point, we have shown that there exists

$p \in L^2(0,T;H_0^1(\Omega)) \cap L^\infty(0,T;L^2(\Omega)) \cap BV([0,T];H^{-s}(\Omega))$ and $\mu_p \in M(Q)$ which are the limit in the sense of (2.24), (2.25), (2.26) and (2.27) on some subsequence (again denoted ε) and satisfy the system

$$(2.28) \qquad p_t - Ap - \mu_p = \varkappa \quad \text{in } Q$$

$$p(T) = \nu \quad \text{in } \Omega$$

$$(2.29) \qquad B^* p(t) \in \partial h(u^*(t)) \quad \text{a.e. } t \in]0,T[$$

Here $\varkappa \in L^2(Q)$ and $\nu \in H$ are the weak limits of $\nabla \Phi^\varepsilon(y_\varepsilon)$ and $\nabla \Psi^\varepsilon(y_\varepsilon(T))$ in $L^2(Q)$ and H, respectively.

We need the following lemma.

LEMMA 2. <u>Let</u> y_n <u>be a sequence strongly convergent to</u> y <u>in</u> H <u>and such that</u>

$$(2.30) \qquad \nabla \Phi^\varepsilon(y_n) \longrightarrow \chi \quad \text{weakly in H for} \quad \varepsilon = n^{-1} \longrightarrow 0.$$

<u>Then</u> $\chi \in \partial \Phi(y)$.

<u>Proof</u>. By the theorem of the mean and formula (2.7) we see that

$$\lambda^{-1}(\Phi^\varepsilon(y_n + \lambda z) - \Phi^\varepsilon(y_n)) = \lambda^{-1}(\Phi(P_n(y_n + \lambda z) - \varepsilon \Lambda_n \tau_{n,\lambda}) -$$

$$- \Phi(P_n y_n - \varepsilon \Lambda_n \tau_{n,\lambda}))$$

where $|\tau_{n,\lambda}|_n \leq 1$. On some subsequence $\lambda \longrightarrow 0$ we have $\tau_{n,\lambda} \longrightarrow \tau_n$ and therefore

$$(\nabla \Phi^\varepsilon(y_n), P_n z) \leq \Phi^0(P_n y_n - \varepsilon \Lambda_n \tau_n, P_n z).$$

Inasmuch as the function Φ^0 is upper semicontinuous on $H \times H$ (see 8) the latter yields $(\chi, z) \leq \Phi^0(y,z)$ for all $z \in H$. Hence $\chi \in \partial \Phi(y)$ as claimed.

Coming back to system (2.28) we observe by (2.31) and Lemma 2 that $\nu \in \partial\Psi(y^*(T))$.

On the other hand, since Φ is locally Lipschitz on H it follows by (2.7) that for every $r > o$ there exists C(r) independent of ε such that

$$\sup\{|\nabla\Phi^\varepsilon(y)| ; |y| \le r\} \le C(r) \quad \text{for all } \varepsilon > o.$$

Hence

$$(2.31) \quad \sup\{|\nabla\Phi^\varepsilon(y_\varepsilon(t))| ; t \in [o,T]\} \le C.$$

Thus by (2.30) we may infer that

$$\nabla\Phi^\varepsilon(y_\varepsilon) \longrightarrow \varkappa \quad \text{weak star in } L^\infty(o,T;H).$$

Since the space H is separable the latter implies that on some subsequence $\varepsilon \longrightarrow o$ we have

$$\nabla\Phi^\varepsilon(y_\varepsilon(t)) \longrightarrow \varkappa(t) \quad \text{weakly in H} \quad \text{a.e.} t \in]o,T[$$

and by Lemma 2 we conclude that

$$\varkappa(t) \in \partial\Phi(y^*(t)) \quad \text{a.e. } t \in]o,T[.$$

We have therefore proved

PROPOSITION 1. Let $(y^*,u^*) \in H^{2,1}(Q) \times L^2(o,T;U)$ be an optimal pair for problem (2.1). Then there exists $p \in BV([o,T];H^{-s}(\Omega)) \cap L^\infty(o,T;H) \cap L^2(o,T;H_o^1(\Omega))$ and $\mu_p \in M(Q)$ such that

$$(2.32) \quad p_t - Ap - \mu_p \in L^\infty(o,T;H)$$

$$(2.33) \quad p_t - Ap - \mu_p \in \partial\Phi(y^*) \quad \text{a.e. on }]o,T[$$

$$(2.34) \quad p(T) + \partial\Psi(y^*(T)) \ni o$$

$$(2.35) \quad B^* p(t) \in \partial h(u^*(t)) \quad \text{a.e. } t \in]o,T[.$$

The function p is the dual extremal arc associated with

the optimal pair (y^*,u^*).

Let us assume now that β satisfies the condition:

(a) β <u>is monotonically increasing, locally Lipschitzian,</u> $\beta(o) = o$ <u>and</u>

(2.36) $\beta'(r) \leq C(|\beta(r)| + |r| + 1)$ a.e. $r \in R$.

PROPOSITION 2. <u>If</u> β <u>satisfies condition</u> (a) <u>then</u> $\mu_p \in L^1(Q) \cap L^1(o,T;H^{-s}(\Omega))$, $p \in AC([o,T];H^{-s}(\Omega))$ <u>where</u> $s > N/2$ <u>and</u>

(2.37) $\mu_p(x,t) \in p(x,t)\partial\beta(y^*(x,t))$, a.e. $(x,t) \in Q$.

<u>Proof</u>. By (2.36) we have

(2.38) $$\int_E |p_\varepsilon| \nabla\beta^\varepsilon(y_\varepsilon)dxdt \leq C\int_E |p_\varepsilon|\,|\beta^\varepsilon(y_\varepsilon)|dxdt +$$

$$+\int_E \beta^\varepsilon(y_\varepsilon)y_\varepsilon\, dxdt + C\int_E |p_\varepsilon| dxdt$$

where E is an arbitrary measurable subset of Q. By Lemma 1 $\beta^\varepsilon(y_\varepsilon)p_\varepsilon$ is weakly convergent while y_ε and p_ε are strongly convergent in $L^2(Q)$. Hence the integrals $\int_E \beta^\varepsilon(y_\varepsilon)\ |p_\varepsilon| dxdt$ and $\int_E \beta^\varepsilon(y_\varepsilon)y\, dxdt$ are uniformly absolutely continuous and by (2.38) we may conclude that the family $\{\int_{E_o} |p_\varepsilon|\nabla\beta^\varepsilon(y_\varepsilon)dt; E_o \subset [o,T]\} \subset L^1(o,T;L^1(\Omega)) \subset L^1(o,T;H^{-s}(\Omega))$ is uniformly absolutely continuous and bounded. Then by the Dunford-Pettis criterion we may infer that $p_\varepsilon\nabla\beta^\varepsilon(y_\varepsilon)$ is weakly compact in $L^1(Q) \cap L^1(o,T;H^{-s}(\Omega))$. Hence $\mu_p \in L^1(Q) \cap L^1(o,T;H^{-s}(\Omega))$. Since $p_t \in L^2(o,T;H^{-s}(\Omega))$, $p \in AC([o,T];H^{-s}(\Omega))$. Formula (2.37) has been proved in [3].

<u>Remark</u> 1° If β is locally Lipschitz and n=1 then $H^{2,1}(Q) \subset C(\bar{Q})$ and by (2.27) we see that $\mu_p \in L^2(Q)$. This also happens if β is globally Lipschitz. Then by (2.18) it follows that $p \in C([o,T];H) \cap L^2(\delta,T;H^2(\Omega))$, $p_t \in L^2(\delta,T;H)$ for every

$o < \delta < T$ (If $p(T) = o$ we may take $\delta = o$.)

3. Optimal feedback control for problem (2.1).

Let $\varphi : [o,T] \times H \longrightarrow R$ be the value function of problem (2.1) defined by (1.10). As remarked in section 1, φ is everywhere finite and the infimum defining $\varphi(t,y_o)$ is attained for every $(t,y_o) \in [o,T] \times H$.

LEMMA 3. <u>For each</u> $t \in [o,T]$ <u>the function</u> $\varphi(t,\cdot)$ <u>is locally Lipschitz and for each</u> $y_o \in D(F)$ <u>the function</u> $t \longrightarrow \varphi(t,y_o)$ <u>is Lipschitz on</u> $[o,T]$.

<u>Proof</u>. Let t be arbitrary but fix in $[o,T]$. By Eq.(1.1) we have

$$(3.1) \qquad |y(s,t,y_o,u)-y(s,t,\tilde{y}_o,u)| \leq |y_o-\tilde{y}_o| \qquad o \leq s \leq T$$

$$(3.2) \qquad |y(s,t,y_o,u)| \leq |y_o| + \|B\| \int_t^s \|u(\tau)\| d\tau \quad , \quad t \leq s \leq T$$

Let $y_o \in H$ be such that $|y_o| \leq r$. We have

$$(3.3) \qquad \varphi(t,y_o) = \inf \{ \int_t^T (\Phi(y(s,o,y_o,u)) + h(u(s)))ds + \Psi(y(T,o,y_o,u))$$

$$; u \in L^2(t,T;U)\} \leq \int_t^T (\Phi(y(s,o,y_o,o)) + h(o))ds +$$

$$+ \Psi(y(T,o,y_o,o)) \leq C(r+1)$$

where C is some positive constant. By assumption 5° we may therefore restrict in (3.3) to all $u \in L^2(t,T;U)$ which satisfy the condition

$$(3.4) \qquad \int_t^T \|u(s)\|^2 ds \leq C_r$$

Let us denote by $\mathcal{M}$ this subset of $L^2(t,T;U)$. By (3.1), (3.2) and (3.4) we see that for $u \in \mathcal{M}$ the function $y_o \longrightarrow \Phi(y(s,t,y_o,u))$ is locally Lipschitz on H with Lipschitz constant (on every bounded subset of H) independent of u. Thus the function $y \longrightarrow \varphi(t,y)$ is locally Lipschitz on H. Next for $y_o \in D(F)$ we have

$$|y(s,t,y_o,u)-y(s,\tilde{t},y_o,u)| \leq |y(t,\tilde{t},y_o,u)-y_o| + \int_t^{\tilde{t}} (\|Bu\| + |Fy_o|)d\tau \quad \text{for } \tilde{t} \leq t \leq s \leq T.$$

Along with the above estimates this implies that $t \longrightarrow \varphi(t,y_o)$ is Lipschitz on $[o,T]$ as claimed.

LEMMA 4. <u>For all</u> $t \in [o,T]$ <u>and</u> $y_o \in H$ <u>we have</u>

$$\varphi(o,y_o) = \inf\{ \int_o^t (\Phi(y(s,o,y_o,u))+h(u(s)))ds + \varphi(t,y(t,o,y_o,u));u \in L^2(o,t;U)\}. \tag{3.5}$$

<u>Proof</u>. Let (y,u) be such that $y(s,o,y_o,u)=y$ and

$$\varphi(o,y_o) = \int_o^t (\Phi(y(s)) + h(u(s)))ds + \int_t^T (\Phi(y(s))+h(u(s)))ds+ \Psi(y(T)).$$

This yields

$$\varphi(o,y_o) \geq \varphi(t,y(t)) + \int_o^t (\Phi(y(s))+h(u(s)))ds. \tag{3.6}$$

On the other hand, for all $u \in L^2(o,T;U)$ and $y=y(s,o,y_o,u)$ we have

$$\varphi(o,y_o) \leq \int_o^t (\Phi(y(s))+h(u(s)))ds +$$

$$+\int_t^T (\Phi(y(s))+h(u(s)))ds + \Psi(y(T)).$$

We may choose the pair (y,u) such that

$$\varphi(t,y(t)) = \int_t^T (\Phi(y(s)) + h(u(s)))ds+ \Psi(y(T))$$

and therefore

$$\varphi(o,y_o) \leq \int_o^t (\Phi(y(s))+h(u(s)))ds+ \varphi(t,y(t)).$$

Along with (3.6) the latter inequality implies (3.5) as claimed.

THEOREM 1. _Assume that conditions_ $1^o \sim 5^o$ _are satisfied. Suppose in addition that_ β _satisfies condition_ (a) _in section_ (2), $\overline{R(B)} = H$ _and_ h _is Gâteaux differentiable. Let_ (y^*,u^*) _be any optimal pair for problem_ (2.1) _where_ y_o _satisfies_ (2.2). _Then_

(3.7) $u^*(t)\in \partial h^*(-B^*\partial \varphi(t,y^*(t))$ a.e. $t\in]o,T[$

Moreover, if p _is the dual extremal arc associated with_ (y^*,u^*) _then_

(3.8) $p(t) +\partial \varphi(t,y^*(t)) \ni o \quad \forall\ t\in [o,T].$

Here $\partial \varphi(t,y)$ is the generalized gradient of the function $y \longrightarrow \varphi(t,y)$.

Proof of Theorem 1. By Lemma 3, for every $t\in [o,T]$ the pair (y^*,u^*) is optimal for the problem

(3.9) $$\inf\{ \int_o^t (\Phi(y(s,o,y_o,u) + h(u(s)))ds + \\ + \varphi(t,y(t,o,y_o,u));u\in L^2(o,t;U)\}.$$

Then in virtue of Proposition 2, for each $t\in [o,T]$ there exists $p^t\in AC([o,t];H^{-s}(\Omega))\cap L^2(o,t;H^1_o(\Omega))\cap L^\infty(o,t;L^2(\Omega))$ such that

(3.10) $B^* p^t(s) = \partial h(u^*(s))$ a.e. $s\in]o,t[$

(3.11) $p^t(t) \in -\partial \varphi(t,y^*(t)).$

Let $p \in AC([o,T];H^{-s}(\Omega))$ be the dual extremal arc coresponding to u^* on $[o,T]$. We have

$$B^* p(s) = \partial h(u^*(s)) \qquad \text{a.e. } s \in]o,t[$$

and by (3.10) we conclude that $p(s) = p^t(s)$ for $s \in [o,t]$ because $N(B^*) = \{o\}$ and the functions p, p^t are continuous on $[o,t]$ with values in $H^{-s}(\Omega)$. Along with (3.11) this implies (3.8) and consequently (3.7) as claimed.

Let us assume now that the following two additional conditions are satisfied

(b) β <u>is locally Lipschitz and either</u> $n=1$ <u>or</u> β <u>is globally Lipschitz</u>.

(c) $\Psi \equiv o$, $h < \infty$ <u>on</u> H, h^* <u>is Fréchet differentiable and</u> ∇h^* <u>is locally Lipschitz on</u> H.

Let y_o be arbitrary in $D(F)$ and let $t \in [o,T]$ be such that $s \longrightarrow \varphi(s,y_o)$ is differentiable at $s = t$. Let $(y^*,u^*) \in H^{2,1}(Q_t) \times L^2(t,T;U)$ be such that $y^*(s) = y(s,t,y_o,u^*)$ for $t \leq s \leq T$ and

$$\varphi(t,y_o) = \int_t^T (\Phi(y^*(s)) + h(u^*(s)))ds. \tag{3.12}$$

By Lemma 3 it follows that

$$\varphi(s,y^*(s)) = \int_s^T (\Phi(y^*(\tau)) + h(u^*(\tau)))d\tau, \quad t \leq s \leq T. \tag{3.13}$$

As remarked at the end of section 2, condition (b) implies that the dual extremal arc p associated with (y^*,u^*) belongs to $W^{1,2}(t,T;H)$. Recalling that

$$u^*(s) = \nabla h^*(B^* p(s)) \qquad \text{a.e. } s \in]t,T[\tag{3.14}$$

it follows from assumption (c) that $u^* \in W^{1,2}(t,T;U)$. This implies (see for instance [1] p.133) that $\frac{d^+}{ds} y^*(s)$ exists everywhere on $[t,T]$ and therefore by (3.13)

$$\text{(3.15)} \quad \frac{d^+}{ds} \varphi(s,y^*(s)) + \Phi(y^*(s)) + h(u^*(s)) = o \quad \text{for all} \quad s \in [t,T].$$

On the other hand, we have

$$\text{(3.16)} \quad \frac{d^+}{ds} \varphi(t,y^*(t)) = \varphi_t(t,y_o) + \lim_{\varepsilon \to o} (\varphi(t+\varepsilon, y^*(t+\varepsilon)) -$$

$$- \varphi(t,y^*(t)))/\varepsilon = \varphi_t(t,y_o) + \lim_{\varepsilon \to o} (\zeta_\varepsilon, y^*(t+\varepsilon) - y^*(t))/\varepsilon$$

where $\zeta_\varepsilon(t) \in \partial\varphi(t+\varepsilon, \theta_\varepsilon)$ and θ_ε is a point in the open line segment between $y_\varepsilon(t)$ and $y_\varepsilon(t+\varepsilon)$ (Here we have used a mean value theorem due to G.Lebourg [10].) Then if we impose the condition :

(d) <u>The</u> map $(t,y) \longrightarrow \partial\varphi(t,y)$ <u>is upper semicontinuous</u> it follows by (1.1)' and (3.15)

$$\text{(3.17)} \quad \varphi_t(t,y_o) + (\zeta(t,y_o), B^* \nabla h^*(-B^*\eta(t,y_o)) + Fy_o) + \Phi(y_o) +$$

$$+ h(\nabla h^*(-B^*\eta(t,y_o))) = o$$

where

$$\zeta(t,y_o) \in \partial\varphi(t,y_o); \; \eta(t,y_o) \in \partial\varphi(t,y_o).$$

Since for $y_o \in D(F)$ the function $s \longrightarrow \varphi(s,y_o)$ is a.e. differentiable on $]o,T[$, it follows that Eqs.(3.17), (3.18) hold for all $y_o \in D(F)$. In this sense we may view φ as solution to the Hamilton-Jacobi equation

$$\text{(3.18)} \quad \varphi_t(t,y) + h(\nabla h^*(-B^*\partial\varphi(t,y))) + (\partial\varphi(t,y), B^*\nabla h^*(-B^*\partial\varphi(t,y)) +$$

$$+ Fy) + \Phi(y) = o; \; y \in D(F), \text{ a.e. } t \in]o,T[$$

with the Cauchy condition

(3.19) $\varphi(T,y) = 0, \quad y \in H.$

Summarising, we have proved

PROPOSITION 3. *Assume that assumptions* $1^\circ \sim 5^\circ$ *and* (b),(c), (d) *are satisfied. Further assume that* h *is Gâteaux differentiable. Then the value function* φ *is the solution to the Cauchy problem* (3.18), (3.19).

If $\partial\varphi$ happens to be single valued at (t,y) then Eq.(3.18) can be written as

$$\varphi_t(t,y) - h^*(-B^*\partial\varphi(t,y)) - (Fy, \partial\varphi(t,y)) + \Phi(y) = 0. \tag{3.20}$$

Remark. Condition (d) is in particular satisfied if $\varphi(t,y)$ is convex in y or if φ admits a (jointly) continuous derivative φ_y. The latter case occurs when the functions Φ , h and β are continuously differentiable and the interval [o,T] is sufficiently small. A direct treatement of Hamilton-Jacobi equation in this case carried out in [5], reveals that φ is of class C^1 on $[0,T] \times H$. The case Φ convex and F linear has been treated in [4] p.293.

4. A control problem with infinite time horizon

We shall study here the problem

$$\inf\left\{ \int_0^\infty (\Phi(y(s,0,y_0,u)) + h(u(s)))ds ; u \in L^2_{loc}(R^+;U)\right\} = \varphi(y_0) \tag{4.1}$$

where $y(s,0,y_0,u)$ is the solution to system (1.1) and Φ ,h, β and $B:U \longrightarrow H$ satisfy the conditions:

(i) $\Phi : H \longrightarrow R^+$ *is locally Lipschitz and* $\Phi(0) = 0$.

(ii) $h: U \longrightarrow R^+$ *is convex, Gâteaux differentiable and*

$$h(u) \geqslant h(0) = 0 \quad \text{for all } u \in U \tag{4.2}$$

(4.3) $h(u) \geq C_1\|u\|^2+C_2$ for all $u \in U$; $C_1 > o$.

(iii) β <u>satisfies condition</u> (a) <u>in section</u> 2 <u>and</u> $\overline{R(B)}=H$.

Observe that for every $y_o \in H$ there exists $u \in L^2(R^+;U)$ such that $\Phi(y(t,o,y_o))+h(u(t)) \in L^1(R^+)$. Indeed it suffices to take $u=-B^*y$ where y is the solution to

$$y'(t)+Fy(t)+BB^*y(t) = o \qquad \text{a.e. } t>o$$
$$y(o) = y_o .$$

By (1.1)' and (1.8) we have $|y(t)| \leq C\exp(-\omega t)|y_o|$ for $t \geq o$ and hence our assumptions imply that $\Phi(y)+h(B^*y) \in L^1(R^+)$. Moreover, as in the proof of Lemma 3 it follows that the function φ is locally Lipschitz on H and for every $y_o \in H$ the infimum in (4.1) is attained in at least one pair $(y^*,u^*) \in H^{2,1}_{loc}(Q^o) \times L^2_{loc}(R^+;U)$.

THEOREM 2. <u>Let assumptions</u> (2.2) <u>and</u> (i), (ii), (iii) <u>be satisfied</u>. <u>If</u> (y^*,u^*) <u>is an optimal pair of problem</u> (4.1) <u>then there exists</u> $p \in L^\infty(R^+;H) \cap L^2_{loc}(R^+;H^1_o(\Omega)) \cap AC_{loc}(R^+;H^{-s}(\Omega))$ $(s > N/2)$ <u>such that</u>

(4.4) $p_t - Ap - p\,\partial\beta(y^*) - \partial\Phi(y^*) \ni o$ in Q^o

(4.5) $B^*p(t) = \partial h(u^*(t))$ a.e. $t>o$

(4.6) $p(t)+\partial\varphi(y^*(t)) \ni o$ for all $t \geq o$.

<u>Proof</u>. Arguing as in the proof of Lemma 3 it follows that for every $t \geq o$,

(4.7) $$\varphi(y_o)=\inf\Big\{\int_o^t (\Phi(y(s,o,y_o,u))+h(u(s)))ds+\varphi(y(t));\ u \in L^2(o,T;U)\Big\}.$$

From Proposition 2 it follows that there exists

$p^t \in L^\infty(o,T;H) \cap L^2(o,t;H_o^1(\Omega)) \cap AC([o,t];H^{-s}(\Omega))$ such that

(4.8) $\quad p_t^t - Ap^t - p^t \partial\beta(y^*) - \partial\Phi(y^*) \ni o \quad$ in Q_t

(4.9) $\quad p^t(t) + \partial\varphi(y^*(t)) \ni o$

(4.10) $\quad B^* p^t(s) = \nabla h(u^*(s)) \qquad$ a.e. $s \in]o,t[$.

Since $N(B^*) = \{o\}$ and ∂h is single valued we see by (4.10)

(4.11) $\quad p^t(s) = p^{t'}(s) \qquad$ for $o \leq s \leq t \leq t'$.

Let $p:R^+ \longrightarrow H$ be the function defined by

(4.12) $\quad p(s) = p^t(s) \qquad$ for $s \in [o,T]$

which obviously satisfies Eqs.(4.4), (4.5), (4.6). Since $h(u^*) \in L^1(R^+)$ it follows by (4.3) that $u^* \in L^2(R^+;U)+L^\infty(R^+;U)$. It follows by $(1.1)'$ and (1.8) that $|y^*(t)|$ is bounded on R^+.Inasmuch as $\partial\varphi$ is locally bounded we see by (4.6) that $p \in L^\infty(R^+;H)$ there by completing the proof.

In particular, it follows by Theorem 2 that the set $\{(y,p) \in H \times H; p+\partial\varphi(y) \ni o\}$ is an <u>invariant manifold</u> of the Hamiltonian system

$$(4.13) \quad \begin{aligned} &y_t + Ay + \beta(y) - B^*\partial h^*(B^*p) \ni o \\ &p_t - Ap - p\partial\beta(y) - \partial\Phi(y) \ni o. \end{aligned}$$

For related results in the case of linear systems of the form (1.1) and Φ convex we refer to [3] (see also [14]).

By Theorem we may also infer that

(4.14) $\quad u \in \partial h^*(-B^*\partial\varphi(y))$

is an optimal feedback law for problem (4.1). By (4.7) we have

$$\varphi(y^*(t)) = \int_t^\infty (\Phi(y^*(s)) + h(u^*(s)))ds \quad \text{for all } t \geq o$$

and therefore

$$\frac{d}{dt}\varphi(y^*(t)) + \Phi(y^*(t)) + h(u^*(t)) = o, \quad \text{a.e. } t > o.$$

Along with (4.5), (4.6) the latter yields

$$(4.15) \quad (\zeta(t), Fy^*(t)) - (B^*\zeta(t), \partial h^*(-B^*\eta(t))) = \Phi(y^*(t)) + \\ + h(\partial h^*(-B^*\eta(t))), \quad \text{a.e. } t > o.$$

Thus there exists a dense subset $E \subset D(F)$ such that

$$(4.16) \quad (\zeta(y_o), Fy_o) - (B^*\zeta(y_o), \partial h^*(-B^*\eta(y_o)) = \Phi(y_o) + \\ + h(\partial h^*(-B^*\eta(y_o))), \quad y_o \in E$$

where $\zeta(y_o)$, $\eta(y_o) \in \partial\varphi(y_o)$ for all $y_o \in H$.

Thus φ can be regarded as a generalized solution to the Hamilton-Jacobi equation (the Bellman equation)

$$(4.17) \quad (\partial\varphi(y), Fy) - (B^*\partial\varphi(y), \partial h^*(-B^*\partial\varphi(y)) = \\ = \Phi(y) + h(\partial h^*(-B^*\partial\varphi(y))).$$

If $\partial\varphi$ is single valued at y then (4.17) becomes

$$(4.18) \quad (\partial\varphi(y), Fy) + h^*(-B^*\partial\varphi(y)) = \Phi(y).$$

For related results in convex case see [6].

<u>Remarks</u> 1° If $h(u) \geqslant C\|u\|^2$ for all $u \in U$ then $\lim_{t \to \infty} y^*(t) = o$ strongly in H.

2° If β satisfies assumption (b) then as observed earlier, the dual extremal arc p in Eq.(4.4) is locally absolutely continuous. Thus if Φ happens to be Fréchet differentiable, it follows by $(1.1)'$ and (4.4),

$$\frac{d}{dt}(p(t), Fy^*(t)) = \frac{d}{dt}(h^*(p(t)) - \Phi(y^*(t))) \quad \text{a.e. } t > o$$

and therefore

$$(4.19) \quad (p(t), Fy^*(t)) - h^*(p(t)) + \Phi(y^*(t)) = C, \quad t \geqslant o.$$

5. Applications to time optimal control problem

Consider the problem

$$(5.1)\quad \inf\{T; y'(t)+Fy(t)=u(t); |u(t)|\leq 1 \quad \text{a.e. } t\in]o,T[\quad y(o)=y_o; y(T)=o\} = T(y_o)$$

where F is defined by (1.6), (1.7), y_o is a fixed element of H, A is a linear elliptic operator satisfying condition 1° in section 1 and β satisfies condition (iii) in section 4.

The value $T(y_o)$ of problem (5.1) is called the optimal time corresponding to y_o and it is easy to see that there exists a control u^* such that $y(T(y_o),o,y_o,u^*)=o$. Such a control is called time optimal control for system (1.1). For linear systems there exist a number of significant results on this problem (see for instance [9] and [12].) Here we shall use a different approach which relies on section 4.

Let $\pi\in C^\infty(R^+)$ be defined by

$$(5.2)\quad \pi(r) = \begin{cases} 1 & \text{for } r\geq 2 \\ o & \text{for } o\leq r\leq 1 \end{cases}$$

and $g^\varepsilon: H \longrightarrow R$ given by

$$(5.3)\quad g^\varepsilon(y) = \pi(|y|^2/\varepsilon^2), \quad y\in H.$$

We set

$$(5.4)\quad G^\varepsilon = \nabla g^\varepsilon; h^\varepsilon(u)=(2\varepsilon)^{-1}((|u|-1)^+)^2 \quad \text{for } u\in H$$

and define the function $\varphi^\varepsilon: H \longrightarrow R$,

$$(5.5)\quad \varphi^\varepsilon(y_o) = \inf\{\int_o^\infty (g^\varepsilon(y(s,o,y_o,u))+h^\varepsilon(u(s)))ds; \; u\in L^2_{loc}(R^+;H)\}$$

where $y(s,o,y_o,u)$ denotes as above the solution to (1.1),((1.1)') where $B\equiv I$. Let us assume that $y_o\in H^1_o(\Omega)$ satisfies condition

(2.2). Thus conditions (i) ~ (iv) are satisfied with $\Phi = g^{\varepsilon}$ and $h = h^{\varepsilon}$ and for every $\varepsilon > o$ problem (5.5) has at least one solution $(y^{\varepsilon}, u^{\varepsilon}) \in H^{2,1}_{loc}(Q^o) \times L^2_{loc}(R^+;H)$. Next by Theorem 2 there exists $p^{\varepsilon} \in L^2_{loc}(R^+;H^1_o(\Omega)) \cap L^{\infty}(R^+;H) \cap AC_{loc}(R^+;H^{-s}(\Omega))$ such that

(5.6)
$$y^{\varepsilon}_t + Fy^{\varepsilon}(t) = u^{\varepsilon} \quad \text{in } Q^o$$
$$p^{\varepsilon}_t - Ap^{\varepsilon} - p^{\varepsilon}\partial\beta(y^{\varepsilon}) \ni G^{\varepsilon}(y^{\varepsilon}) \quad \text{in } Q^o$$

(5.7)
$$p^{\varepsilon}(t) = \nabla h^{\varepsilon}(u^{\varepsilon}(t)) = \begin{cases} \dfrac{u^{\varepsilon}(t)}{|u^{\varepsilon}(t)|}(|u^{\varepsilon}(t)| - 1) & \text{if } |u^{\varepsilon}(t)| \geqslant 1 \\ o & \text{if } |u^{\varepsilon}(t)| < 1 \end{cases}$$

(5.8)
$$p^{\varepsilon}(t) + \partial\varphi^{\varepsilon}(y^{\varepsilon}(t)) \ni o \quad \text{for } t \geqslant o.$$

By (5.4) and (5.7) it follows that

(5.9)
$$(h^{\varepsilon})^*(p) = |p| + \frac{\varepsilon}{2}|p|^2 \quad , \quad p \in H$$

and therefore

(5.10)
$$u^{\varepsilon}(t) = -\text{sgn}\, p^{\varepsilon}(t) - \varepsilon\, p^{\varepsilon}(t) \quad \text{a.e. } t > o.$$

Along with (5.8) the latter implies that

(5.11)
$$u(t) = \text{sgn}\,\partial\varphi^{\varepsilon}(y(t)) - \varepsilon\partial\varphi^{\varepsilon}(y(t)) \; , \quad t > o$$

is an <u>optimal feedback law</u> for problem (5.5).

By (4.16) and (5.9) we see that for every $\varepsilon > o$, φ^{ε} is the solution to the stationary Hamilton-Jacobi equation

(5.12)
$$\frac{\varepsilon}{2}|\partial\varphi^{\varepsilon}(y)|^2 + |\partial\varphi^{\varepsilon}(y)| + (Fy, \partial\varphi^{\varepsilon}(y)) = g^{\varepsilon}(y)$$

i.e., there exist $\zeta_{\varepsilon} \in \partial\varphi^{\varepsilon}$ and $\eta_{\varepsilon} \in \partial\varphi^{\varepsilon}$ such that

(5.13)
$$(\zeta_{\varepsilon}(y), Fy + \text{sgn}\,\eta_{\varepsilon}(y) + \varepsilon\eta_{\varepsilon}(y)) = g^{\varepsilon}(y), \quad y \in E \subset D(F).$$

Now let us assume that β <u>satisfies assumption</u> (b) in section 4. Then (see Eq.(4.19)) we have

(5.14) $(p^\varepsilon(t),Fy^\varepsilon(t))-|p^\varepsilon(t)|-\varepsilon|p^\varepsilon(t)|^2+g^\varepsilon(y^\varepsilon(t))=0$ a.e. $t>0$

($C=0$ because $|p^\varepsilon|=\varepsilon^{-1}(|u^\varepsilon|-1)^+\in L^2(R^+)$ and $g^\varepsilon(y^\varepsilon)\in L^1(R^+)$.) Recalling that $p^\varepsilon(t)\in-\partial\varphi^\varepsilon(y^\varepsilon(t))$ we see that in this case φ^ε satisfies Eq.(5.12) in a stronger sense, i.e., $\zeta_\varepsilon=\eta_\varepsilon$.

The relevance of function φ^ε in the time optimal control problem (5.1) is explained in Theorem 3 below.

THEOREM 3. <u>Let</u> y_0 <u>be any element of</u> H <u>satisfying condition</u> (2.2). <u>Then</u> $\lim_{\varepsilon\to 0}\varphi^\varepsilon(y_0)=T(y_0)$ <u>and on some subsequence we have</u>

(5.15) $u^\varepsilon\longrightarrow u^*$ <u>weakly in</u> $L^2_{loc}(R^+;H)$

(5.16) $y^\varepsilon\longrightarrow y^*$ <u>strongly in every</u> $C([0,T];H)$

<u>where</u> u^* <u>is a time optimal control and</u> $y^*(t)=y(t,0,y_0,u^*)$ <u>is the corresponding state</u>.

<u>Proof</u>. Let $T^*=T(y_0)$ be the optimal time and let (y_1^*,u_1^*) be any optimal pair for problem (5.1). We have

(5.17) $$\varphi^\varepsilon(y_0)\le\int_0^\infty(g^\varepsilon(y_1^*(t))+h^\varepsilon(u_1^*(t)))dt=$$
$$=\int_0^{T^*}(g^\varepsilon(y_1^*(t))+h^\varepsilon(u_1^*(t)))dt.$$

Thus

(5.18) $$\limsup_{\varepsilon\to 0}\varphi^\varepsilon(y_0)\le T^*.$$

In particular it follows that $\{u^\varepsilon\}$ is bounded in every $L^2(0,T;H)$ and $\{y^\varepsilon\}$ is compact in every $C([0,T];H)$. Hence there exist $(y^*,u^*)\in H^{2,1}_{loc}(Q^0)\times L^2_{loc}(R^+;H)$ such that $y^*(t)=y(t,0,y_0,u^*)$ and

$u^{\varepsilon} \longrightarrow u^{*}$ weakly in $L^2_{loc}(R^+;H)$

$y^{\varepsilon}(t) \longrightarrow y^{*}(t)$ uniformly on every $[o,T]$.

Next by (5.18)

$$\limsup_{\varepsilon \to o} \int_o^T g^{\varepsilon}(y^{\varepsilon}(t))dt \leq T^{*} \quad \text{for every } T > o. \tag{5.19}$$

It follows that there exist a sequence $\varepsilon_n \longrightarrow o$ and T_o independent of n such that

$$|y^{\varepsilon_n}(t)| \leq 2\varepsilon_n \quad \text{for} \quad t \geq T_o. \tag{5.20}$$

For, otherwise for every sequence ε_n convergent to zero would exist $t_n \longrightarrow \infty$ such that $|y^{\varepsilon_n}(t_n)| > 2\varepsilon_n$ for all n. Let $\varepsilon_n = n^{-1/2}$. Then by an easy calculation involving Eq.(1.1)' it would follow that

$$|y^{\varepsilon_n}(t_n)| \leq |y^{\varepsilon_n}(t)| + \int_t^{t_n} |u^{\varepsilon_n}(s)|\, ds \quad \text{for} \quad t \leq t_n. \tag{5.21}$$

Since $(2\varepsilon_n)^{-1} \int_o^{\infty} ((|u^{\varepsilon_n}(t)| - 1)^+)^2 dt \leq C$, by (5.21) it follows that

$$|y^{\varepsilon_n}(t_n)| \leq |y^{\varepsilon_n}(t)| + |t - t_n| + C(2\varepsilon_n |t - t_n|)^{1/2}, \; t \leq t_n$$

and therefore

$$|y^{\varepsilon_n}(t)| \geq \sqrt{2}\, \varepsilon_n \quad \text{for} \quad t \in [t_n - \delta_n, t_n]$$

where $\delta_n = C_o n^{-1}$ (C_o is a positive constant). This would imply that $\lim_{n \to \infty} m\{t; |y^{\varepsilon_n}(t)| \geq \sqrt{2}\, \varepsilon_n\} = +\infty$ (m is the Lebesgue measure), contrary to (5.19).

By (5.20) it follows that $y^{*}(t) = o$ for $t \geq T_o$. Let $\tilde{T} = \inf\{T; y^{*}(T) = o\}$. We will show that $\tilde{T} = T^{*}$. To this end for $\varepsilon > o$ consider the set $E_{\varepsilon} = \{t \in [o, \tilde{T}]; |y^{\varepsilon}(t)| \geq \sqrt{2\varepsilon}\}$. By (5.19),

(5.22) $\limsup_{\varepsilon \to 0} m(E_\varepsilon) \leq T^*$.

On the other hand, $\limsup_{\varepsilon \to 0} m(E_\varepsilon) = \tilde{T}$ for, otherwise would exist $\delta > 0$ and $\varepsilon_n \longrightarrow 0$ such that $m(E_{\varepsilon_n}) \leq \tilde{T} - \delta$. In other words, would exist a sequence of measurable subsets $A_n \subset [0,T]$ such that $m(A_n) \geq \delta$ and $|y^{\varepsilon_n}(t)| \geq \sqrt{2}\,\varepsilon_n$ for $t \in A_n$. This would imply that $|y^*(t)| \leq \sqrt{2}\,\varepsilon_n + \delta_n$ for $t \in A_n$ where $\delta_n \longrightarrow 0$. On the other hand, since $y^*(t) \neq 0$ for $t \in [0,\tilde{T}]$, $m(t; |y^*(t)| \leq \sqrt{2}\,\varepsilon_n + \delta_n) \longrightarrow 0$ for $n \longrightarrow \infty$. The contradiction we arrived at shows that indeed $\limsup m(E_\varepsilon) = \tilde{T}$. Along with (5.22) the latter implies that $\tilde{T} = T^*$ as claimed. Thus u^* is a time optimal control and the proof is complete.

<u>Remark</u>. From the preceding proof it is apparent that (5.15) holds for all $y_0 \in H$. Recalling that φ^ε is a solution to Eq.(5.12) we may formally regard $T: H \longrightarrow R$ as a solution to the Bellman equation

$$(5.23)\qquad (Fy, \partial T(y)) + |\partial T(y)| = 1$$

where ∂T is the "gradient" of T in some generalized sense.

The main consequence of Theorem 3 is the fact that (5.11) is an <u>approximating feedback control</u> for problem (5.1).

Now we shall study the following variant of the time optimal problem: minimize

$$(5.24)\qquad \frac{\alpha}{2}\int_0^T |u|^2 dt + T$$

over all $(y,u) \in H^{2,1}(Q) \times L^2(0,T;U)$ subject to

$$(5.25)\qquad \begin{aligned} & y' + Fy = u \qquad \text{a.e. } t \in]0,T[\\ & y(0) = y_0, \qquad y(T) = 0, \end{aligned}$$

where α is a positive constant.

We associate with (5.24) the approximating problem

$$(5.26)\quad \inf\left\{\int_0^\infty (g^\varepsilon(y(t,o,y_o,u)) + \frac{\alpha}{2}|u(t)|^2)dt\right\} = \phi^\varepsilon(y_o).$$

Let $(y_\varepsilon, u_\varepsilon) \in H^{2,1}_{loc}(Q^o) \times L^2(R^+;H)$ be an optimal pair for problem (5.26). By Theorem 2 there exists $p_\varepsilon \in L^2_{loc}(R^+;H^1_o(\Omega)) \cap L^2(R^+;H) \cap L^\infty(R^+;H) \cap AC_{loc}(R^+;H^{-s}(\Omega))$ which satisfies the system

$$(5.27)\quad \begin{aligned} &(y_\varepsilon)_t + Fy_\varepsilon = \alpha^{-1}p_\varepsilon && \text{in } Q^o \\ &(p_\varepsilon)_t - Ap_\varepsilon - p_\varepsilon \partial\beta(y) = G^\varepsilon(y_\varepsilon) && \text{in } Q^o \end{aligned}$$

$$(5.28)\quad p_\varepsilon(t) + \partial\phi^\varepsilon(y_\varepsilon(t)) \ni o \qquad \text{for all } t \geq o$$

$$(5.29)\quad u_\varepsilon = \alpha^{-1}p_\varepsilon.$$

As noticed earlier, $\lim_{t\to\infty} y_\varepsilon(t) = o$ in H. By (4.16) we see that ϕ^ε is the solution to the Hamilton-Jacobi equation

$$(5.30)\quad (Fy, \partial\phi^\varepsilon(y)) + (2\alpha)^{-1}|\partial\phi^\varepsilon(y)|^2 = g^\varepsilon(y)$$

and

$$(5.31)\quad u = -\alpha^{-1}\partial\phi^\varepsilon(y)$$

is an <u>optimal feedback control</u> for problem (5.26).

THEOREM 4. <u>For every</u> $y_o \in H$, $\lim_{\varepsilon\to o} \phi^\varepsilon(y_o) = \phi(y_o)$ <u>and on some sequence</u> $\varepsilon_n \longrightarrow o$

$$(5.32)\quad u_{\varepsilon_n} \longrightarrow u_1^* \quad \text{strongly in } L^2(R^+;H)$$

$$(5.33)\quad y_{\varepsilon_n} \longrightarrow y_1^* \quad \text{uniformly in H on } R^+$$

<u>where</u> (y_1^*, u_1^*) <u>is an optimal pair of problem</u> (5.24).

The proof which is essentially the same as that of Theorem 3 will be omitted.

We may view ϕ as a generalized solution to the Bellman equation

$$(5.34)\qquad (Fy,\partial\psi(y)) + (2\alpha)^{-1}|\partial\psi(y)|^2 = 1, \quad y \neq o.$$

There remains the question whether Eq.(5.34) has solutions in the sense precised above. On these lines it is instructive to notice that if β satisfies condition (b) then by (5.27), (5.29) we have (see (4.19))

$$(5.35)\quad -(Fy_\varepsilon(t),u_\varepsilon(t))+\frac{1}{2}|u_\varepsilon(t)|^2 = \alpha^{-1}g^\varepsilon(y_\varepsilon(t)),\ \text{a.e. } t>o.$$

Arguing as in the proof of Theorem 3 we see that $g^\varepsilon(y_\varepsilon) \longrightarrow 1$ a.e. $t\in]o,T^*[$. Thus letting ε tend to zero in (5.3) we get

$$(5.36)\qquad -(Fy_1^*(t),u_1^*(t)) + \frac{1}{2}|u_1^*(t)|^2 = \alpha^{-1} \quad \text{a.e. } t\in]o,T^*[$$

where T^* is the optimal time in problem (5.24).

To implement a suboptimal feedback control of the form (5.31) it would be desirable to have existence for the Cauchy problem

$$y'+Fy+\partial\psi(y) \ni o \qquad \text{a.e. } t>o;\ y(o) = y_o$$

where ψ is a locally Lipschitz function on H. By the results of [15] it follows that this happens for instance if $(w,y)\geqslant o$ for all $(w,y)\in\partial\psi$.

R E F E R E N C E S

1. V.BARBU, - Nonlinear Semigroups and Differential Equations in Banach Spaces, Noordhoff 1976.

2. V.BARBU, - Convex control problems and Hamiltonian systems on infinite intervals, SIAM <u>J.Control and Optimiz</u>. 16(1978), 687-702.

3. V.BARBU, - Necessary conditions for distributed control problems governed by parabolic variational inequalities, <u>SIAM J.Control and Optimization</u> 19(1981), 64-86.

4. V.BARBU, Th.PRECUPANU, - Convexity and Optimization in Banach Spaces, Noordhoff § Sijthoff 1978.

5. V.BARBU, G.DA PRATO, - Hamilton-Jacobi equations and synthesis of nonlinear control processes in Hilbert space, J.Diff.Equations (to appear).

6. V.BARBU, G.DA PRATO, - Existence and approximation for stationary Hamilton-Jacobi equations, J.Nonlinear Analysis 6(1981).

7. F.H. CLARKE,- Generalized gradients and applications, Trans. Amer. Math. Soc. 205(1975), 247-262.

8. F.H. CLARKE,- Generalized gradients of Lipschitz functionals, Advances in Math. 40(1981), 52-67.

9. H.O. FATTORINI,- The time optimal control problem in Banach space, Applied Math. § Optimiz. Volume 1 (1974), 163-188.

10. G.LEBOURG, - Valeur moyenne pour le gradient généralisé, C.R. Acad. Sci.Paris 281(1975), 795-797.

11. J.L.LIONS, - Quelques méthodes de resolution des problèmes aux limites non lineaires, Dunod Gauthier - Villars Paris 1969.

12. J.L.LIONS, - Optimal Control of Systems Governed by Partial Differential Equations, Springer-Verlag 1971.

13. R.T.ROCKAFELLAR, - Directionally lipschitzian functions and subdifferential calculus, Proc.London Math.Soc. 39(1979), 331-355.

14. R.T.ROCKAFELLAR, - Saddle points of Hamiltonian systems in convex problem of Lagrange, J.Optimiz. Theory Appl. 12(1973), 367-390.

15. I.VRABIE, - The nonlinear version of Pazy's local existence theorem, Israel J.Math. 32(1979), 221-235.

ON THE PRODUCTION SMOOTHING PROBLEM

A. BENSOUSSAN *

INTRODUCTION

We discuss in this paper the problem of production smoothing considered by KUNREUTHER - MORTON [2]. The K.M. model is in discrete time. In this framework we generalize the assumptions of K.M. and give a more rigorous proof of their results. We also present a continuous time version of the problem, which is new. Many results extend, but unfortunately the planning horizon result which is obtained in the discrete time case does not extend easily to the continuous time case.

1. THE DISCRETE TIME PRODUCTION SMOOTHING PROBLEM

1.1. The model

We consider the following problem in discrete time

$$(1.1)\qquad \left|\begin{array}{ll} y_{k+1} = y_k + v_k - \xi_{k+1} & k = 0,\dots,N-1 \\ y_0 = x & \end{array}\right.$$

$$(1.2)\qquad v_k \geq 0 \quad , \quad y_k \geq 0 \qquad\qquad k = 0,\dots,N$$

$$(1.3)\qquad J_0^N(x;V) = \sum_{j=0}^{N-1} c_j(v_j) + \sum_{j=1}^{N-1} f_j(y_j) + \sum_{j=1}^{N-1} h_j(v_j - v_{j-1}) + h_N(-v_{N-1})$$

with the following assumptions

(1.4) $c_i(v)$, $f_i(x)$, $i = 0,\dots,N-1$ are functions from R^+ into R^+, which are non decreasing and convex, and not reduced to a constant.

(*) University Paris Dauphine and INRIA.

(1.5) $h_i(z) : R \to R^+$ convex, and achieves its minimum at 0.

The control variables are $V = (v_0, v_1, \ldots)$ and represent amounts to be produced. The state variables are $y_0, y_1, \ldots$ and represent inventories. The functions $c_j(v_j)$ are ordering costs, the functions $f_j(x)$ are inventory costs, and the functions $h_j(z)$ are penalty costs to avoid big variations of the rate of production. The last cost $h_N(-v_{N-1})$ tends to diminish v_{N-1} (if $h_N \not\equiv 0$).

The sequence $\xi_1, \xi_2, \ldots, \xi_N$ of positive values represents the demand to be satisfied. It is convenient to consider V as an infinite sequence with the constraint

(1.6) $$v_N = 0$$

and to define

(1.7) $$c_N(v) = 0 \quad , \quad f_N(x) = 0 \quad , \quad h_0(z) = 0 \ .$$

Our objective is to study the properties of optimal controls (which exist, but many be not unique) and to adress the problem of the planning horizon.

We call problem (1.3) the N horizon problem. We note

$$J^N(V) = J_0^N(0;V) \ .$$

We say that N is a *planning horizon,* if for any $K > N$, we can obtain an optimal policy for problem $J^K(V)$ by taking an optimal control $\hat{v}_0^N, \ldots, \hat{v}_{N-1}^N$ of the N horizon problem and completing it adequately at times $N, \ldots, K-1$.

Planning horizons are important in practice since they provide decisions which are robust with respect to the horizon (they remain the same as the horizon increases).

1.2. Necessary conditions of optimality

We can reformulate (1.1), (1.2), (1.3), (1.6) in a way which is more standard from the point of view of control theory, namely find

$$y_0, y_1, \ldots, y_N, v_0, \ldots, v_N, w_0, \ldots, w_{N-1}$$

satisfying

(1.8)
$$\left|\begin{array}{ll} y_{k+1} = y_k + v_k - \xi_{k+1} , & k = 0,\dots,N \\ v_{k+1} = v_k + w_k & \\ y_0 = x , \quad v_N = 0 & \\ y_k, v_k \geq 0 & \end{array}\right.$$

to minimize

(1.9)
$$J_0 = \sum_{j=0}^{N-1} c_j(v_j) + \sum_{j=1}^{N-1} f_j(y_j) + \sum_{j=0}^{N-1} h_{j+1}(w_j) .$$

Let $\hat{y}_k$, $\hat{v}_k$, $\hat{w}_k$ be an optimal solution, there exist variables λ_k^1, $k = 1,\dots,N$ λ_k^2, $k = 0,\dots,N-1$, and p_k^1, p_k^2, $k = 0,\dots,N-1$, such that

(1.10)
$$\lambda_k^1 , \lambda_k^2 \geq 0 \qquad \hat{y}_0 = x , \quad \hat{v}_N = 0$$

$$\lambda_k^1 \hat{y}_k = 0 , \quad k = 1,\dots,N$$

$$\lambda_k^2 \hat{v}_k = 0 , \quad k = 0,\dots,N-1$$

$$\hat{y}_{k+1} = \hat{y}_k + \hat{v}_k - \xi_{k+1} , \quad k = 0,\dots,N-1$$

(1.11)
$$p_{N-1}^1 - \lambda_N^1 = 0$$

(1.12)
$$h'^-_{k+1}(\hat{v}_{k+1}-\hat{v}_k) \leq p_k^2 \leq h'^+_{k+1}(\hat{v}_{k+1}-\hat{v}_k) , \quad k = 0,\dots,N-1$$

(1.13)
$$f'^-_k(\hat{y}_k) \leq p_k^1 - p_{k-1}^1 + \lambda_k^1 \leq f'^+_k(\hat{y}_k) , \quad k = 1,\dots,N-1$$

(1.14)
$$c'^-_k(\hat{v}_k) \leq p_k^1 + p_k^2 - p_{k-1}^2 + \lambda_k^2 \leq c'^+_k(\hat{v}_k) , \quad k = 1,\dots,N-1$$

(1.15)
$$c'^-_0(\hat{v}_0) \leq p_0^1 + p_0^2 + \lambda_0^2 \leq c'^+_0(\hat{v}_0) .$$

These conditions form a set of necessary and sufficient conditions of optimality. They are obtained by standard methods of convex analysis.

2. PROPERTIES OF OPTIMAL CONTROL

We shall in the sequel assume $x = 0$, without mentionning it. Our first objective is to characterize the *regeneration points* of an optimal control. Those are points k such that $\hat{y}_k = 0$.

In the classical production problem without smoothing (particular case corresponding to $h_j = 0$), regeneration points play a drastic role in the theory of planning horizons. In fact, regeneration points which are stable with respect to the horizon are planning horizons. Although such properties do no carry over to the case of smoothing, it remains important to identify regeneration points.

2.1. Preliminary results

We state the following

Lemma 2.1. Assume that

(2.1) $$\forall\, k \le N-1$$

$$c_k'^{+}(0) + \sum_{j=k+1}^{N} f_j'^{+}(0) - h_{k+1}'^{-}(0) > -h_k'^{+}(-\infty)$$

then an optimal control satisfies

(2.2) $$\hat{y}_N = 0\ .$$

Proof

Assume that $\hat{y}_N > 0$. There exists $k_0 \leqslant N-1$ (the last set up) such that

$$\hat{v}_{k_0} > 0 \quad , \quad \hat{y}_{k_0+1} > 0 \ \dots\ -\hat{y}_N > 0 \quad , \quad \hat{v}_{k_0+1} = \dots = \hat{v}_N = 0$$

hence

$$\lambda_{k_0}^2 = 0 \quad , \quad \lambda_{k_0+1}^1 = \dots = \lambda_N^1 = 0\ .$$

We thus have from (1.14)

$$c_{k_0}'^{-}(\hat{v}_{k_0}) \le p_{k_0}^1 + p_{k_0}^2 - p_{k_0-1}^2$$

and from (1.13)

$$\sum_{j=k_0+1}^{N} f_j'^-(\hat{y}_j) \le -p_{k_0}^1 \quad .$$

Adding up

$$c_{k_0}'^-(\hat{v}_{k_0}) + \sum_{j=k_0+1}^{N} f_j'^-(\hat{y}_j) \le p_{k_0}^2 - p_{k_0-1}^2$$

and from (1.12)

$$\le h_{k_0+1}'^+(\hat{v}_{k_0+1} - \hat{v}_{k_0}) - h_{k_0}'^-(\hat{v}_{k_0} - \hat{v}_{k_0-1})$$

therefore also since $\hat{v}_{k_0+1} - \hat{v}_{k_0} < 0$

$$c_{k_0}'^+(0) + \sum_{j=k_0+1}^{N} f_j'^+(0) \le h_{k_0+1}'^-(0) - h_{k_0}'^+(-\infty)$$

which contradicts (2.1). □

<u>Example</u>. Let us consider the following example

(2.3) $$c_j(v) = 0 \quad , \quad h_j(z) = \frac{\alpha}{2}|z| \quad , \quad f_j(x) = x$$

then (2.1) is clearly satisfied. □

The next question deals with the maximum length of time which separates two successive regeneration points. More precisely, for any k define $H^*(k) \ge 1$ to be the first interger ≥ 1 such that

(2.4) $$\sum_{j=k+1}^{k+H^*(k)} (j-k)\, f_j'^+(0) + \sum_{j=k+H^*(k)+1}^{k+2H^*(k)} (k+2H^*+1-j)\, f_j'^+(0) +$$

$$\sum_{j=k}^{k+H^*-1} c_j'^+(0) > \sum_{j=k+H^*+1}^{k+2H^*} c_j'^-(\infty) + h_{k+H^*}'^-(+\infty) + h_{k+H^*+1}'^-(+\infty) -$$

$$h_h'^+(-\infty) - h_{k+2H^*+1}'^+(-\infty) \; .$$

In the example, $H^*(k) = H^*$ which is the first interger ≥ 1 such that

(2.5) $$H^*(H^*+1) > 2\alpha \; .$$

The result is the following

Lemma 2.2. Assume (2.1) and

$$c_k'^{+}(0) + f_{k+1}'^{+}(0) - 2h_{k+1}'^{-}(0) > c_{k+1}'^{-}(+\infty) - h_k'^{+}(-\infty) - h_{k+2}'^{+}(-\infty) \tag{2.6}$$

$$\forall\ k \leq N-2$$

Then if $0 \leq k \leq N-1$ *is a regeneration point of an optimal policy (i.e.* $\hat{y}_k = 0$ *), there exists* ℓ *with* $k+1 \leq \ell \leq (k+2H^*(k)) \wedge N$ *which is also a regeneration point.*

Assumption (4.6) is clearly satisfied in the model example.

The proof relies on the necessary conditions and adequate modification of the optimal control. Details are messy and omitted (see A. Bensoussan - M. Crouhy - J.M. Proth [1]).

Remark 2.1. Assume (2.6), for some $k \leq N-2$. Then one has

$$\hat{v}_k > \hat{v}_{k+1} \quad \text{implies} \quad \hat{y}_{k+1} = 0\ . \tag{2.7}$$

Indeed if $\hat{y}_{k+1} > 0$ then $\lambda_k^2 = 0 \quad , \quad \lambda_{k+1}^1 = 0$

$$c_k'^{-}(\hat{v}_k) \leq p_k^1 + p_k^2 - p_{k-1}^2$$

$$f_{k+1}'^{-}(\hat{y}_{k+1}) \leq p_{k+1}^1 - p_k^1$$

$$p_{k+1}^1 + p_{k+1}^2 - p_k^2 \leq c_{k+1}'^{+}(\hat{v}_k)$$

Adding up

$$c_k'^{-}(\hat{v}_k) + f_{k+1}'^{-}(\hat{y}_{k+1}) + p_{k+1}^2 + p_{k-1}^2 \leq 2p_k^2 + c_{k+1}'^{+}(\hat{v}_k)\ .$$

Using (1.12), we deduce easily

$$c_k'^{+}(0) + f_{k+1}'^{+}(0) - 2h_{k+1}'^{-}(0) \leq c_{k+1}'^{-}(\infty) - h_k'^{+}(-\infty) - h_{k+2}'^{+}(-\infty)$$

which contradicts (2.6). □

We next give an estimate on the size of orders

Lemma 2.3. Assume (2.1), (2.6) and

(2.8) $H^*(k)$ *does not decrease as* k *increases.*

Then one has $(\forall\ k \le N-1)$

(2.9) $$\hat{v}_k \le D_k^N = \underset{1\le t\le 2H^*(k)\wedge(N-k)}{\text{Max}} \frac{\sum_{j=1}^{t} \xi_{k+j}}{t}$$

Proof

By virtue of (2.8), one can check that $\forall\ k \le N-1$, there will be ℓ with $k \le \ell \le (k+2H^*(k)) \wedge N$ such that $\hat{y}_\ell = 0$, whether $\hat{y}_k$ vanishes or not. Assuming ℓ to be the smallest such integer, one has

$$\hat{y}_{k+1} > 0 \quad \dots \quad , \quad \hat{y}_{\ell-1} > 0$$

hence

$$\hat{v}_k + \dots + \hat{v}_{\ell-1} \le \xi_{k+1} + \dots + \xi_\ell .$$

But from Remark 2.1, we have

$$\hat{v}_k \le \hat{v}_{k+1} \dots \le \hat{v}_{\ell-1}$$

hence

$$\hat{v}_k \le \frac{\xi_{k+1} + \dots + \xi_\ell}{\ell-k} \le D_k^N$$ □

Denote by $\hat{v}_j^k$, $j \le k-1$ the jth component of an optimal control for the problem with horizon k. We shall consider in the sequel the assumptions

(2.10) $$h_j(x) = h_j(0) - \bar{h}_j\, x \quad \text{for} \quad x \le 0$$

$$\bar{h}_j \ge 0 \quad , \quad \bar{h}_j \ge \bar{h}_{j+1}$$

(2.11) $$c_j'^-(+\infty) < f_k'^+(0) \quad \forall\ j \ge k$$

Those assumptions imply (2.1) and (2.6).

2.2. The main results

We first have the following important lemma

Lemma 2.4. We assume (2.8), (2.10), (2.11). Then if

$$\hat{v}_{k-1}^{k} > D_k^N \tag{2.12}$$

one has

$$\hat{v}_{k-1}^{k} > \hat{v}_k \tag{2.13}$$

and there exists an optimal control which contains $\hat{v}_j^k$, $j = 0,\dots,k-1$.

Proof

Property (2.13) follows directly from (2.12) and (2.9). Let us prove the second part. Assume first that

$$\hat{y}_k = 0 \ . \tag{2.14}$$

Define

$$\tilde{J}^k(V) = \sum_{j=1}^{k-1} h_j(v_j-v_{j-1}) + \sum_{j=0}^{k-1} c_j(v_j) + \sum_{j=1}^{k-1} f_j(y_j) \tag{2.15}$$

$$L^k(V;w) = \tilde{J}^k(V) + h_k(w-v_{k-1}) \tag{2.16}$$

$$J^k(V) = \tilde{J}^k(V) + h_k(-v_{k-1}) = L^k(V;0) \ .$$

Because of (2.14) the optimal control $\hat{V}$ minimizes $L^k(V;\hat{v}_k)$ among controls such that $y_k = 0$. Since $\hat{V}^k$ is admissible, we have

$$\begin{aligned} L^k(\hat{V};\hat{v}_k) &= \tilde{J}^k(\hat{V}) + h_k(\hat{v}_k-\hat{v}_{k-1}) \\ &\le \tilde{J}^k(\hat{V}^k) + h_k(\hat{v}_k-\hat{v}_{k-1}^k) \end{aligned} \tag{2.17}$$

On the other hand, since $\hat{V}^k$ minimizes $J^k(V)$ we also have

$$\tilde{J}^k(\hat{V}^k) + h_k(-\hat{v}_{k-1}^k) \le \tilde{J}^k(\hat{V}) + h_k(-\hat{v}_{k-1}) \ . \tag{2.18}$$

Combining (2.17), (2.18) we deduce

(2.19) $$h_k(-\hat{v}^k_{k-1}) - h_k(-\hat{v}_{k-1}) \le h_k(\hat{v}_k-\hat{v}^k_{k-1}) - h_k(\hat{v}_k-\hat{v}_{k-1}) .$$

Let us prove also the reverse inequality. Indeed if

$$\hat{v}^k_{k-1} \ge \hat{v}_{k-1}$$

then, by convexity the reverse inequality holds.

If

$$\hat{v}^k_{k-1} \le \hat{v}_{k-1}$$

by (2.13) one has

$$\hat{v}_k < \hat{v}_{k-1}$$

and using (2.10)

$$h_k(-\hat{v}_{k-1}) - h_k(-\hat{v}^k_{k-1}) - h_k(\hat{v}_k-\hat{v}_{k-1}) + h_k(\hat{v}_k-\hat{v}^k_{k-1}) =$$

$$= \bar{h}_k[\hat{v}_{k-1} - \hat{v}^k_{k-1} - (\hat{v}_{k-1}-\hat{v}_k) + \hat{v}^k_{k-1} - \hat{v}_k] = 0 .$$

Since (2.19) is thus an equality, we deduce from (2.17), (2.18) that

$$L^k(\hat{V};\hat{v}_k) = L^k(\hat{V}^k;\hat{v}_k)$$

which is sufficient to conclude the desired result.

Assume next that

(2.20) $$\hat{y}_k > 0 .$$

We shall show that this contradicts the optimality of $\hat{V}$, hence cannot occur. Indeed, if (2.20) holds then by Remark 2.1 and (2.13)

$$\hat{v}_{k-1} \le \hat{v}_k < \hat{v}^k_{k-1} .$$

There exists ℓ with $k+1 \le \ell \le (k+2H^*(k)) \wedge N$ such that $\hat{y}_\ell = 0$. Let ℓ^* with $k+1 \le \ell^* \le \ell$ be the smallest integer such that

(2.21) $$\hat{y}_{\ell^*} \le (\ell^*-k)\, \hat{v}^k_{k-1} - \sum_{j=k+1}^{\ell^*} \xi_j .$$

Note that by (2.12), this inequality is satisfied by the index ℓ.

One defines a new control as follows

(2.22)
$$\tilde{v}_j = \hat{v}^k_j \ , \ 0 \le j \le k-1 \ , \ \tilde{v}_j = \hat{v}^k_{k-1} \ , \ k \le j \le \ell^*-2$$
$$\tilde{v}_{\ell^*-1} = \hat{y}_{\ell^*} + \sum_{j=k+1}^{\ell^*} \xi_j - (\ell^*-1-k)\,\hat{v}^k_{k-1} \ , \ \hat{v}_j = \tilde{v}_j \ , j \ge \ell^*$$

Note that $\tilde{v}_{\ell^*-1} \ge 0$. The following sequence of inventories corresponds to $\tilde{V}$

(2.23)
$$\begin{cases} \tilde{y}_j = \hat{y}^k_j \ , & 0 \le j \le k \\ \tilde{y}_j = (j-k)\hat{v}^k_{k-1} - (\xi_{k+1}+\dots+\xi_j) \ , & k+1 \le j \le \ell^*-1 \\ \tilde{y}_j = \hat{y}_j \ , & \ell^* \le j \le N \ . \end{cases}$$

From the definition of ℓ^* and Remark 2.1 we can assert that

(2.24)
$$\hat{v}_{k-1} \le \hat{v}_k \dots \le \hat{v}_{\ell^*-1} \le \tilde{v}_{\ell^*-1} \le \hat{v}^k_{k-1}$$

Using (2.18) and $\tilde{y}_j \le \hat{y}_j$ for $j = k+1,\dots,\ell^*-1$ we obtain

(2.25)
$$\begin{aligned} J(\tilde{V}) - J(\hat{V}) \le\ & h_k(-\hat{v}_{k-1}) - h_k(-\hat{v}^k_{k-1}) + h_k(\tilde{v}_{\ell^*-1}-\hat{v}^k_{k-1}) + \\ & + h_{\ell^*}(\hat{v}_{\ell^*}-\tilde{v}_{\ell^*-1}) - h_{\ell^*}(\hat{v}_{\ell^*}-\hat{v}_{\ell^*-1}) - \\ & - \sum_{j=k}^{\ell^*-1} h_j(\hat{v}_j-\hat{v}_{j-1}) + \sum_{j=k}^{\ell^*-2} (c_j(\hat{v}^k_{k-1})-c_j(\hat{v}_j)) + \\ & + c_{\ell^*-1}(\tilde{v}_{\ell^*-1}) - c_{\ell^*-1}(\hat{v}_{\ell-1}) + f_k(0) - f_k(\hat{y}_k) \ . \end{aligned}$$

We use (2.10), (2.24), to deduce

$$\begin{aligned} J(\tilde{V}) - J(\hat{V}) \le\ & \bar{h}_k(\hat{v}_{k-1}-\tilde{v}_{\ell^*-1}) + h_{\ell^*}(\hat{v}_{\ell^*}-\tilde{v}_{\ell^*-1}) - h_{\ell^*}(\hat{v}_{\ell^*}-\hat{v}_{\ell^*-1}) - \\ & - \sum_{j=k}^{\ell^*-1} h_j(\hat{v}_j-\hat{v}_{j-1}) + \sum_{j=k}^{\ell^*-2} (c_j(\hat{v}^k_{k-1}) - c_j(\hat{v}_j)) + \\ & + c_{\ell^*-1}(\tilde{v}_{\ell^*-1}) - c_{\ell^*-1}(\hat{v}_{\ell^*-1}) + f_k(0) - f_k(\hat{y}_k) \ . \end{aligned}$$

From the definition of $\tilde{v}_{\ell^*-1}$ we also have

$$\hat{y}_k = \tilde{v}_{\ell^*-1} - \hat{v}_{\ell^*-1} + \sum_{j=k}^{\ell^*-2} (\hat{v}^k_{k-1} - \hat{v}_j) \ .$$

By convexity

$$c_j(\hat{v}^k_{k-1}) - c_j(\hat{v}_j) \le c'_j(\hat{v}^k_{k-1})(\hat{v}^k_{k-1}-\hat{v}_j)$$

$$\le c'_j(\infty)\ (\hat{v}^k_{k-1}-\hat{v}_j) \quad , \quad j = k,\dots,\ell^*-2$$

$$c_{\ell^*-1}(\tilde{v}_{\ell^*-1}) - c_{\ell^*-1}(\hat{v}_{\ell^*-1}) \le c'_{\ell^*-1}(\infty)\ (\tilde{v}_{\ell^*-1}-\hat{v}_{\ell^*-1})$$

$$f_k(\hat{y}_k) - f_k(0) \ge f'^{+}_k(0)\ \hat{y}_k$$

$$= f'^{+}_k(0)\ [\tilde{v}_{\ell^*-1} - \hat{v}_{\ell^*-1} + \sum_{j=k}^{\ell^*-2} (\hat{v}^k_{k-1}-\hat{v}_j)]$$

hence

$$\sum_{j=k}^{\ell^*-2} (c_j(\hat{v}^k_{k-1})-c_j(\hat{v}_j)) + c_{\ell^*-1}(\tilde{v}_{\ell^*-1}) - c_{\ell^*-1}(\hat{v}_{\ell^*-1}) + f_k(0) - f_k(\hat{y}_k)$$

$$\le (c'_{\ell^*-1}(\infty)-f'^{+}_k(0))(\tilde{v}_{\ell^*-1}-\hat{v}_{\ell^*-1}) + \sum_{j=k}^{\ell^*-2} (c'_j(\infty)-f'^{+}_k(0))(\hat{v}^k_{k-1}-\hat{v}_j)$$

$$< 0 \quad \text{by (2.11)} .$$

Therefore we have

$$J(\tilde{V}) - J(\hat{V}) < \bar{h}_k(\hat{v}_{k-1}-\tilde{v}_{\ell^*-1}) + h_{\ell^*}(\hat{v}_{\ell^*}-\tilde{v}_{\ell^*-1}) - h_{\ell^*}(\hat{v}_{\ell^*}-\hat{v}_{\ell^*-1}) -$$

$$- \sum_{j=k}^{\ell^*-1} h_j(\hat{v}_j-\hat{v}_{j-1}) = X$$

Let us prove that $X \le 0$ which will imply the desired contradiction. Consider first the case when $\hat{v}_{\ell^*} \le \hat{v}_{\ell^*-1}$, then also $\hat{v}_{\ell^*} \le \tilde{v}_{\ell^*-1}$ and

$$X = \bar{h}_k(\hat{v}_{k-1}-\tilde{v}_{\ell^*-1}) + \bar{h}_{\ell^*}(\tilde{v}_{\ell^*-1}-\hat{v}_{\ell^*-1}) - \sum_{j=k}^{\ell^*-1} h_j(\hat{v}_j-\hat{v}_{j-1})$$

and by (2.10)

$$\bar{h}_{\ell^*} \le \bar{h}_k$$

hence

$$X \le \bar{h}_k(\hat{v}_{k-1}-\hat{v}_{\ell^*-1}) - \sum_{j=k}^{\ell^*-1} h_j(\hat{v}_j-\hat{v}_{j-1}) \le 0$$

Assume next that $\hat{v}_{\ell^*} \ge \hat{v}_{\ell^*-1}$, and $\hat{v}_{\ell^*} \ge \tilde{v}_{\ell^*-1}$.

Then

$$\hat{v}_{\ell^*}-\hat{v}_{\ell^*-1} \geq \hat{v}_{\ell^*}-\tilde{v}_{\ell^*-1} \geq 0$$

and since h_{ℓ^*} increases with positive arguments, again $X \leq 0$. The last case is $\hat{v}_{\ell^*} \geq \hat{v}_{\ell^*-1}$ and $\hat{v}_{\ell^*} \geq \tilde{v}_{\ell^*-1}$.

Then

$$X \leq \bar{h}_k(\hat{v}_{k-1}-\tilde{v}_{\ell^*-1}) + \bar{h}_{\ell^*}(\tilde{v}_{\ell^*-1}-\hat{v}_{\ell^*})$$

$$\leq \bar{h}_k(\hat{v}_{k-1}-\hat{v}_{\ell^*}) \leq \bar{h}_k(\hat{v}_{\ell^*-1}-\hat{v}_{\ell^*}) \leq 0 \ .$$

The above considerations were valid for $\ell^* \geq k+2$; if $\ell^* = k+1$ a similar and simpler argument holds. The proof has been completed. □

The preceding lemma yields the following planning horizon theorem

Theorem 2.1. Assume (1.4), (1.5), (2.8), (2.10), (2.11). If

(2.26)
$$\hat{v}_{n-1} > D_N = \max_{1 \leq t \leq 2H^*(N)} \frac{\sum_{j=1}^{t} \xi_{N+j}}{t}$$

then N *is a planning horizon.*

Proof

Let $K > N$, (2.26) implies

$$\hat{v}^N_{N-1} > D^K_N$$

But then Lemma 2.4 yields that there exists an optimal control for $J^K(V)$, which contains $\hat{v}^N_j$, for $j = 0,\dots,N-1$. Hence the desired result. □

Remark 2.2. Although by the planning horizon property, the N-1 first decisions do not depend on future demands, one needs to know future demands to decide whether N is a planning horizon or not. In fact one needs $2H^*(N)$ demands beyond N. □

Remark 2.3. Assumption (2.10) together with the last part of (1.5) prevents $h_i(z)$ to be continuously differentiable, at 0. No planning horizon result seems possible without a discontinuity of the derivative at 0. □

3. THE CONTINUOUS TIME PRODUCTION SMOOTHING PROBLEM

3.1. Set up of the problem

Let us consider

(3.1) $\xi(t) \geq 0 \qquad \int_0^T \xi(t)\,dt < \infty$

and functions $c(t;v)$, $f(t;x)$, $h(t;w)$ satisfying

(3.2) $c(t;v)$, $f(t;x)$: $[0,T] \times R^+ \rightarrow R^+$, convex non decreasing (in v,x respectively), Borel in both arguments and $c'^+(t;0) > 0$.

(3.3) $h(t;w)$: $[0,T] \times R \rightarrow R^+$, convex in w, Borel in both arguments, and achieves its minimum at 0.

$g(T;v)$: $R^+ \rightarrow R^+$, convex, non negative, non decreasing.

The dynamic system is described as follows

(3.4)
$$\frac{dy}{dt} = v(t) - \xi(t)$$
$$y(0) = x$$
$$\frac{dv}{dt} = w(t)$$

with the constraints

(3.5) $y(t) \geq 0 \quad , \quad v(t) \geq 0.$

The control is composed of the pair

(3.6) $w(.) \in L^2(0,T) \quad , \quad v(0) \geq 0 .$

We want to minimize the cost function

(3.7)
$$J_0(x;w(.),v(0)) = \int_0^T c(t;v(t))dt + \int_0^T f(t;y(t))dt + \int_0^T h(t;w(t))dt + g(T;v(T)) .$$

This functional is convex, but under the above assumptions we cannot guarantee the existence of an optimal control. We will postulate its existence, and look for necessary and sufficient conditions. We denote by $\hat{w}(.)$, $\hat{y}(.)$, $\hat{v}(.)$ an optimal control and corresponding state. Moreover we assume that

(3.8) $$\int_0^T |h'^{\pm}(t;\hat{w}(t))|^2 dt \quad , \quad \int_0^T |c'^{\pm}(t;\hat{v}(t))|^2 dt \quad ,$$

$$\int_0^T |f'^{\pm}(t;\hat{y}(t))|^2 dt$$

finite .

3.2. Necessary and sufficient conditions of optimality

We have the following

Theorem 3.1. We assume (3.1), (3.2), (3.3). Let $\hat{w}$ *be an optimal control for (3.7), such that (3.8) holds. Then there exist* $\lambda(.)$, $\mu(.)$, $p_1(.)$, $p_2(.)$ *right continuous functions with bounded variations such that*

(3.9) λ,μ *are non decreasing*, $\lambda(0) = \mu(0) = 0$

(3.10) $\lambda+p_1$, $\mu+p_2 \in H^1(0,T)$

$\lambda+p_1 \in c([0,T])$, $\mu+p_2 \in c([0,T])$

(3.11) $p_1(T) = p_2(T) = 0$

(3.12) $\int_0^T \hat{y}(t)d\lambda(t) = 0$

(3.13) $\int_0^T \hat{v}(t)d\mu(t) = 0$

(3.14) $h'^{-}(t;\hat{w}(t)) \le p_2(t) \le h'^{+}(t;\hat{w}(t))$ a.e.

(3.15) $\frac{d}{dt}(\mu(t) + p_2(t)) + p_1(t) \le c'^{+}(t;\hat{v}(t))$ a.e.

$\frac{d}{dt}(\mu(t) + p_2(t)) + p_1(t) \ge c'^{-}(t;\hat{v}(t))$ a.e. (1)

---

(1) $c'^{-}(t;0) = c'^{+}(t;0)$, $g'^{-}(T;0) = g'^{+}(T;0)$ by convention.

(3.16) $$\frac{d}{dt}(\lambda(t)+p_1(t)) \leq f'^{+}(t;\hat{y}(t)) \quad \text{a.e.}$$

$$\frac{d}{dt}(\lambda(t)+p_1(t)) \geq f'^{-}(t;\hat{y}(t)) \quad \text{a.e.}$$

(3.17) $$p_2(0) = 0$$

(3.18) $$g'^{-}(T;\hat{v}(T)) \leq (p_2+\mu)(T) - (p_2+\mu)(T^-) \leq g'^{+}(T;\hat{v}(T))$$ □

The proof is omitted and follows from the theory of convex optimization in infinite dimensional spaces.

Our objective is to see what are the results of the discrete time case which carry over to the present framework.

3.2. Properties of optimal controls

We assume now

(3.19) $$x < \int_0^T \xi(t)dt$$

(3.20) $$\forall\ 0 < s < s_1 < T\ ,$$

$$h'^{-}(s_1;0) - h'^{+}(s;-\infty) < \int_s^{s_1} c'^{+}(t;0) + \int_s^{s_1} dt \int_t^T f'^{+}(\lambda;0)d\lambda$$

$$\forall\ 0 < s < T\ ,$$

$$-g'^{+}(T;0) - h'^{+}(s;-\infty) < \int_s^T c'^{+}(t;0)dt + \int_s^T dt \int_t^T f'^{+}(\lambda;0)d\lambda$$

We state

Theorem 3.2. We make the assumptions of Theorem 3.1 and (3.19), (3.20). Then an optimal control for $J(w(.),v(0))$ *satisfies*

$$\hat{y}(T) = 0\ .$$

Remark 3.1. (3.20) is an integrated form of the continuous analogue of (2.1). It is restrictive with respect to the derivatives $h'^{-}(t;x)$, $h'^{+}(t;x)$ for $x < 0$. It is satisfied when for instance

$$h(t;x) = -\bar{h}(t)x \quad , \quad \text{for} \quad x < 0$$

and $\bar{h}$ is increasing with $\bar{h}(T) < g'^{+}(T;0)$. □

The next step is to derive a result similar to that of Lemma 2.2.

We define for any s, $H^*(s)$ by the formula

$$\begin{aligned}(3.21) \qquad H^*(s) = \inf_{\theta>0}\{\theta \int_s^{s+2\theta} f'^{+}(\tau;0)d\tau + \int_s^{s+\theta}(\tau-s)\, f'^{+}(\tau;0)d\tau + \\ + \int_{s+\theta}^{s+2\theta} (s+\theta-\tau)f'^{+}(\tau;0)d\tau + \int_s^{s+\theta} c'^{+}(\tau;0)d\tau > \\ > \int_{s+\theta}^{s+2\theta} c'^{-}(\tau;\infty)d\tau + 2h'^{-}(s+\theta;+\infty) - h'^{+}(s;-\infty) - \\ - h'^{+}(s+2\theta;-\infty)\}\end{aligned}$$

We need the following assumption

(3.22) $\quad \forall\ s < s_1 < t$

$$\begin{aligned}\frac{1}{s_1-s} \int_s^{s_1} c'^{+}(\tau;0)d\tau + \frac{1}{t-s_1} \int_{s_1}^{t} (t-\tau)f'^{+}(\tau;0)d\tau + \\ + \frac{1}{s_1-s} \int_s^{s_1}(\tau-s)\, f'^{+}(\tau;0)d\tau > \frac{1}{t-s_1} \int_{s_1}^{t} c'^{-}(\tau;+\infty)d\tau + \\ + \frac{1}{s_1-s}\,(h'^{-}(s_1;0)-h'^{+}(s;-\infty)) + \frac{1}{t-s_1}\,(h'^{-}(s_1;0) - h'^{+}(t;-\infty))\end{aligned}$$

We then have the following result

Theorem 3.3. We make the assumptions of Theorem 3.1 and (3.19), (3.20), (3.22). Assume that for some s *an optimal control satisfies* $\hat{y}(s) = 0$. *Then there exists a point* s^* *in* $(s,(s+2H^*(s)) \wedge T]$ *such that* $\hat{y}(s^*) = 0$. □

We can finally state the analogue of Lemma 2.3.

Theorem 3.4. We make the assumptions of Theorem 3.3 and

(3.23) $\qquad H^*(s)$ *is non decreasing.*

Then, if $x = 0$, *we have*

$$\hat{v}(s) \leq D_s^T \quad , \quad \forall\, s < T \tag{3.24}$$

$$D_s^T = \sup_{0<\theta<2H^*(s)\wedge(T-s)} \frac{1}{\theta} \int_s^{s+\theta} \xi(\lambda)d\lambda \tag{3.25}$$ □

Unfortunately we do not know if a result similar to that of Theorem 2.1 is true. One can conjecture that if $\hat{v}^s(s) \geq D_s^T$, then there is an optimal control which contains $\hat{v}^s(t)$, $t \leq s$, where $\hat{v}^s(.)$, $\hat{w}^s(.)$, $\hat{y}^s(.)$ denotes the optimal solution for the s horizon problem.

REFERENCES

[1] A. Bensoussan - M. Crouhy - J.M. Proth, Production Management, North Holland, to be published.

[2] H.C. Kunreuther - T.E. Morton , General Planning Horizons for Production Smoothing with Deterministic Demand, I, II, Management Science, Vol. 20 (1973), pp. 110-125 et Vol. 20 (1974), pp. 1037-1047.

Existence of solutions and existence of optimal solutions

Lamberto Cesari

A lecture at the International Conference on MATHEMATICAL THEORIES OF OPTIMIZATION at S. Margherita Ligure, Nov. 30-Dec. 4, 1981.

In this lecture we present some novel use of fixed point theorems leading to existence theorems for abstract equations, and in particular for quasi linear evolution equations, and to existence theorems for optimal solutions of corresponding problems of optimization.

1. The topological statement

We present first results which in a sense continue the line of work on fixed point theorems of multifunctions initiated by S. Kakutani [15] (1940) and then developed at great length by many authors as, to name a few, H. F. Bohnenblush and S. Karlin [3] (1950), I. Glicksberg [11] (1952), K. Fan [10] (1952), I. Tarnove [22] (1967), N. Kenmochi [17] (1971).

Here is a theorem which essentially contains some of their statements:

(1.i) (A fixed point theorem). Let K be a closed convex set in a locally convex topological Hausdorff space, let $\Gamma: K \to K$ be a multifunction such that $\Gamma(x)$ is a closed and convex subset of K for every $x \in K$; assume that Γ has a closed graph in $K \times K$, and that $\bigcup [\Gamma(x),\ x \in K]$ is contained in a compact set. Then, Γ has at least one fixed point, that is, there is some $x \in K$ such that $x \in \Gamma(x)$.

For work on fixed point theorems for stochastic maps we mention here Kannan [16ab]. In terms of an abstract equation

(1.1) $$Ex + Nx = 0,\ x \in K,$$

here is a result which is a particularization of a Kenmochi's theorem, proved by this author for multifunctions. Only for the sake of simplicity we restate it here for single valued maps. We need recall a few well known definitions. A multifunction $N: D(N) \subset X \to X^*$ from a Banach space X into its dual X^* is said to be monotone provided $x_1, x_2 \in D(N)$, $y_1 \in Nx_1$, $y_2 \in Nx_2$ implies $\langle y_1 - y_2, x_1 - x_2 \rangle \geq 0$ where $\langle\ ,\ \rangle$ denotes the pairing of X and X^*. Any such multifunction N is said to be maximal monotone provided N is monotone and in addition the graph of N in $X \times X^*$ cannot be properly enlarged so as to remain the graph of a monotone map (G. J. Minty [20], H. Brezis [2c]). Of course, if E is a single valued linear map it is enough for monotonicity to verify that $\langle Ex,x \rangle \geq 0$ for all $x \in D(E)$.

(1.ii) (N. Kenmochi [17], th.2, p. 438). Let K be a bounded closed convex subset of a real reflexive Banach space X containing the origin as an interior point, let $E: D(E) \subset X \to X^*$ be a linear maximal monotone operator, let $N: X \to X^*$ be any bounded map of the Brezis type (m), and assume that $\langle Nx,x \rangle \geq 0$ for all $x \in \partial K$, the boundary of K. Then equation (1.1) has at least one solution in K.

Here ∂K denotes the boundary of K. Also, the map N is said to be bounded if it maps bounded sets into bounded sets, and N is said to be of Brezis type (m) provided

(m_1) If $[x_i]$ is a net such that $\|x_i\| \leq C$, $x_i \rightharpoonup x$ in X, $Nx_i \rightharpoonup x^*$ in X^*, then $Nx = x^*$; and

(m_2) The restriction of N to any finite dimensional subspace of X is continuous with respect to weak topology.

Note that here X and therefore X^* are reflexive spaces, and hence their weak and weak star topologies coincide.

The following statement is a corollary of (1.ii).

(1.iii) (S. H. Hou [14d]). Let K be a bounded closed convex subset of a real reflexive Banach space X containing the origin as an interior point, let $E: D(E) \subset X \to X^*$ be a linear maximal monotone operator with domain D(E), and let $N: X \to X^*$ be an operator, not necessarily linear with domain D(N) = X, and sequentially weakly continuous from X to X^*. Assume that $\langle Nx,x \rangle \geq 0$ for all $x \in \partial K$. Then equation (1.1) has at least one solution in K.

The proof is relevant here. First let us prove that N is bounded. Indeed, in the opposite case, there would be a bounded sequence $[b_s]$ with $[Nb_s]$ unbounded. We could extract a subsequence, say still $[b_s]$, with $|Nb_s|_{X^*} \to \infty$, and since X is reflexive we could extract a further subsequence, say still $[b_s]$, with b_s weakly convergent in X. Then $[Nb_s]$ would be also weakly convergent and hence bounded, a contradiction. Finally, let us prove that N is an operator of the Brezis type (m), namely, let us prove (m_1) and (m_2). Let $[x_i]$ as in (m_1). Then the closed ball $B = [x \in X, |x|_X \leq C]$ is weakly compact because X is reflexive. Therefore, the weak topology of B can be metrized (see, e.g. [9], th. V, 6.3). Thus, there is a subsequence $[b_n]$ of $[x_i]$, $[b_n] \subset B$, such that $b_n \rightharpoonup x$. Since N is by hypothesis sequentially weakly continuous, we conclude that $Nb_n \rightharpoonup Nx$. Thus, $Nx = x^*$, and (m_1) is proved. Of course, (m_2) is obvious. Now statement (1.iii) is contained in (1.ii).

In particular, under the assumptions of (1.iii) for K and E, if $A: X \to X^*$ is any operator, linear or nonlinear, with domain $D(A) = X$ and sequentially weakly continuous from X to X^*, with $\langle Ax,x \rangle \geq \alpha |x|_X^2$ for all $x \in X$ and some constant $\alpha > 0$, and if f is a fixed element of X^*, then the equation $Ex + Ax = f$ has at least one solution $x \in X$. Indeed, for the nonlinear operator $N: X \to X^*$ defined by $Nx = Ax - f$, and for $|x|_X = R$ we have

$$\langle Nx,x \rangle = \langle Ax - f,x \rangle$$
$$\geq \alpha |x|_X^2 - |f|_{X^*} |x|_X = \alpha R^2 - |f|_{X^*} R;$$

hence, $\langle Nx,x \rangle \geq 0$ for all $|x|_X = R$ and $R \geq \alpha^{-1} |f|_{X^*}$.

2. Trajectories and controls.

Let $J = [0,T]$ be a given interval.

Let V, V^* be dual reflexive Banach spaces, $p,q > 1$, $1/p + 1/q = 1$, and let $X = L_p(J,V)$ denote the space of all functions $x(t)$, $t \in J$, with values in V, and L_p-integrable norm $\|x(t)\|_V$. Let $\|x\|_X = (\int_0^T \|x(t)\|_V^p dt)^{1/p}$ denote the norm of x in X. Then $X = L_p(J,V)$ and $X^* = L_q(J,V^*)$ are dual reflexive spaces.

If $\langle v,w\rangle$ denote the pairing of V and V^*, then the pairing of X and X^* will be $\{x,y\} = \int_0^T \langle x(t),y(t)\rangle\ dt$.

We are concerned here with systems described by an equation of the form

$$(2.1)\qquad \begin{aligned} &(Ex)(t) + (\Lambda x)(t) = g(t,(Mx)(t),u(t)),\ t \in J = [0,T] \\ &u(t) \in \omega(t),\quad x \in X. \end{aligned}$$

Here Λ and M are given operators, and the second member of the equation is a Nemitsky operator. Also, for the sake of generality, we include here the case in which the same second member depends on a control function $u(t)$, $t \in J$, with values in a given control set $\omega(t) \subset U$, which depends only on t in a given space U. In other words $\omega\colon J \to U$ is a given multifunction from J to U. Any measurable selection u of ω will be called a control; any pair $x(t)$, $u(t)$, $t \in J$, $x \in X$, u a control, satisfying (2.1), will be called an admissible pair (x an admissible trajectory, u an admissible control).

Thus, for any control $u(t)$, and $N_u x = \Lambda x(\cdot)-g(\cdot,(Mx)(\cdot),u(\cdot))$, equation (2.1) has the usual form $Ex + N_u x = 0$, $x \in X$. Let Z be a given Banach space and $Y = L_1(J,Z)$. We assume that

(A_1) $E\colon D(E) \subset X \to X^*$ is a linear maximal monotone operator with domain $D(E) \subset X$.

(A_2) $\Lambda\colon X \to X^*$ is an operator with domain $D(\Lambda) = X$, and Λ is sequentially weakly continuous from X to X^*, i.e., if $x_k \to x$ weakly in X, then $\Lambda x_k \rightharpoonup \Lambda x$ in the weak star topology of X^*.

(A_3) $M\colon X \to Y$ is an operator with domain $D(M) = X$, mapping $X = L_p(J,V)$ into $Y = L_1(J,V)$, and M maps weak convergent sequences $[x_k]$ of X into strongly convergent sequences $[Mx_k]$ of Y.

(A_4) There is a bounded closed convex set K of X, containing the origin as an interior point, such that $\langle N_u x, x\rangle \geq 0$ for any measurable selection u of $\omega(t)$ and for any $x \in \partial K$, the boundary of K.

(A_5) The function g(t,z,u) is a map $J \times Z \times U \to V^*$, and for every measurable selection u of ω and for every $z \in Z$ we have $|g(t,z,u(t))|_{V^*} \leq \ell(t)$ for a fixed function $\ell \in L_q(J,\mathbb{R})$. Moreover, for a.a. $t \in J$ and $z_k \to z$ strongly in Z, then $g(t,z_k,u(t)) \longrightarrow g(t,z,u(t))$ weakly in V^* (in other words, the map $z \to g(t,z,u(t))$ is demicontinuous for a.a. $t \in J$).

We are now in a position to prove the following statement.

(2.i) Under conditions (A_1-A_5), and for every measurable selection u of ω, system (2.1) has at least one solution in K.

First we have to prove that $N_u x = \Lambda x - g(\cdot,(Mx),(\cdot)u(\cdot))$ as a map from X to X^* is weakly sequentially continuous. First, let us prove that M is continuous from the weak topology of X to the strong topology of Y. Indeed, if $x_k \longrightarrow x$ in X then, by (A_3), $Mx_k \to y$ strongly in Y for some y. But also the sequence $x_1,x,x_2,x,\cdots,x_k,x,\cdots$ converges weakly to x in X, and again by (A_3) the corresponding sequence in Y must converge strongly to Mx. Thus, $Mx_k \to Mx$ strongly in Y, and y = Mx.

Also, M is bounded from X to Y, and the proof is the same as for Λ in no. 1. Let us prove that, for any selection u of ω, the operator g defined by $gx = g(\cdot,Mx(\cdot),u(\cdot))$ is weakly continuous. Indeed, for every measurable selection u of ω, and sequence $x_k \longrightarrow x$ weakly convergent in X, then by (A_3), $Mx_k \to Mx$ strongly in $Y = L_1(J,Z)$; hence, $(Mx_k)(t) \to (Mx)(t)$ strongly in Z for a.a. $t \in J$. By (A_5) and for a.a. $t \in J$, then $g(t,(Mx_k)(t),u(t)) \longrightarrow g(t,(Mx)(t),u(t))$ weakly in V^*, with $|g(t,(Mx_k)(t),u(t))|_{V^*} \leq \ell(t)$, $\ell \in L_q(J,\mathbb{R})$. Finally, $gx_k \longrightarrow gx$ weakly in $X^* = L_q(J,V^*)$. Since $\Lambda x_k \longrightarrow \Lambda x$ by (A_2), the map

$N_u = \Lambda - g$ has the same property. Statement (2.i) is now a corollary of (1.iii).

In the particular case in which, for a.a. $t \in J$ and all $v \in V$ we have $\langle \Lambda v, v \rangle \geq \alpha |v|_V^p$ for some constant $\alpha > 0$, then for $K = [x \in X, |x|_X \leq R]$ and $x \in \partial K$, then $|x|_X = R$. If u denotes any measurable selection of ω, and $g(t) = g(t,(Mx)(t),u(t))$, then $|x|_X = R$, $g(t) \in V^*$, $|g(t)|_{V^*} \leq \ell(t)$, $\ell \in L_q(J,\mathbb{R})$, and

$$\begin{aligned}\{N_u x, x\} &= \int_0^T \langle \Lambda x(t) - g(t), x(t) \rangle \, dt \\ &\geq \alpha \int_0^T |x(t)|_V^p dt - \int_0^T |g(t)|_{V^*} |x(t)|_V \, dt \\ &\geq \alpha \, |x|_X^p - \|\ell\|_q |x|_X = \alpha R^p - \|\ell\|_q R.\end{aligned}$$

Thus, $\{N_u x, x\} \geq 0$ for $R \geq \left(\alpha^{-1} \|\ell\|_q\right)^{\frac{1}{p-1}}$ and $x \in \partial K$. Statement (2.i) is now a corollary of (1.iii).

3. Weak compactness of the class of solutions.

Let us denote by $\Phi = \{x\}$ the class of all solutions $x \in X$, $x \in K$, of (2.1) for all possible measurable selections u of ω. Since $\Phi \subset K$ and K is a bounded subset of the reflexive Banach space X, then obviously Φ is relatively sequentially weakly compact. Under mild assumptions we can show that Φ is actually sequentially weakly compact in X, and this is an important information since, for instance, the existence of optimal solutions in problems of optimization monitored by equation (2.1) follows rather straightforward (Cf. [5]). Let $Q(t,z)$ denote the subset of V^* defined by

$$Q(t,z) = g(t,z,\omega(t)) = [\zeta = g(t,z,u), u \in \omega(t)] \subset V^*, \quad t \in J, \ z \in Z.$$

We shall assume that the sets $Q(t,z)$ satisfy property (Q) with respect to z in Z, that is, for a.a. $t \in J$ and for every sequence $[z_k]$ strongly convergent to some z in Z we have

$$\bigcap_{k=1}^{\infty} \operatorname{cl} \operatorname{co} \bigcup_{s=k}^{\infty} Q(t,z_s) \subset Q(t,z).$$

This property is often stated in the slightly stronger form

$$\bigcap_{\delta>0} \operatorname{cl} \operatorname{co} \bigcup [Q(t,z'), \ |z'-z|_Z \leq \delta] = Q(t,z).$$

In this form, then the sets $Q(t,z)$ are closed and convex in V^* (Cf. Cesari [Uoh]).

In the situation stated above we have proved in [5] that

(3.i) Under the same assumptions (A_1-A_5), if for a.a. $t \in J$ the subset $Q(t,z)$ of V^* have property (Q) with respect to z in Z, then Φ is a weakly sequentially compact subset of K in X.

Let us consider now the particular but important case where a fixed bounded domain G in the ξ-space R^n is given, $\xi = (\xi_1,\cdots,\xi_n)$, where (2.1) is actually a partial differential system in the cylinder $[0,T] \times G$, where V is the L_p-space of certain m-vector functions on G (and possibly a number of their distributional derivatives), so that actually each element $x \in X = L_p(J,V)$ is an m-vector function $x(t,\xi) = (x^1,\cdots,x^m)$, $(t,\xi) \in G$, with $x \in (L_p(JxG,\mathbb{R}))^m$ (with components possessing a number of their distributional derivatives with respect to $\xi_1,\cdots,\xi_n$ also of class $L_p(JxG,R)$). Then, for $Z = \mathbb{R}^s$, $x \in X$, $Y = L_1(J,\mathbb{R}^s)$, we can think of $(Mx)(t)$ as represented by an s-vector function $y(t,\xi) = (y^1,\cdots,y^s)$, $y \in (L_1(JxG))^s$. In this situation we may take any measurable multifunction $\omega_o(t,\xi)$, or ω_o: $J \times G \to \mathbb{R}^s$, and then $u = u(t,\xi)$ represents any measurable selection $u(t,\xi) \in \omega_o(t,\xi) \subset \mathbb{R}^s$. Here $Ex, \Lambda x, g(\cdot,z,u)$ can be thought of as r-vector functions of class $(L_q(JxG,\mathbb{R}))^r$. Finally, instead of (A_5) we shall assume that

(A_5') The function $g(t,\xi,z,u)$ is a map $J \times G \times \mathbb{R}^s \times \mathbb{R}^m \to \mathbb{R}^r$ which is measurable in (t,ξ) for all (z,u), and continuous in (z,u) for a.a. $(t,\xi) \in J \times G$. Moreover $|g(t,\xi,z,u)| \leq \ell(t)$ for some $\ell \in L_q(J,\mathbb{R})$ and all $(t,\xi,z,u) \in J \times G \times \mathbb{R}^s \times \mathbb{R}^m$.

We shall see in no. 5 a case where $n \geq 1$, $G \subset \mathbb{R}^n$, $m=s=r=1$, $\mu \geq 1$.

Under the above assumptions, instead of the subsets $Q(t,y)$ of V^* we shall consider the subsets $Q^*(t,\xi,z,u)$ of $\mathbb{R}^r$ defined by

$$Q^*(t,\xi,z) = g(t,\xi,z,\omega_o(t,\xi)) = [\zeta = g(t,\xi,z,u),\ u \in \omega_o(t,\xi)] \subset \mathbb{R}^r,$$
$$(t,\xi) \in J \times G.$$

Instead of property (Q) we shall need property (K) with respect to z, that is, we shall assume that for a.a. $(t,\xi) \in J \times G$ and for every sequence $[z_k]$ convergent to some z in $\mathbb{R}^s$, we have

$$\bigcap_{k=1}^{\infty} \mathrm{cl} \bigcup_{s=k}^{\infty} Q^*(t,\xi,z_k) \subset Q^*(t,\xi,z).$$

Propriety (K) is often stated in the slightly stronger form

$$\bigcap_{\delta>0} \mathrm{cl} \bigcup [Q^*(t,\xi,z'), |z'-z|_Z \leq \delta] = Q^*(t,\xi,z).$$

In this form, then the sets Q^* are closed, and property (K) is then equivalent to the requirement that the graph of $Q^*(t,\xi,z)$, as z describes Z, is closed in $Z \times \mathbb{R}^r$ (Cf. Cesari [4ah]).

Under the above assumptions we have proved in [5] the following statement in which the same contention as in (3.i) is obtained under essentially weaker requirements.

(3.ii) (Cesari and Hou [5]) Under the same assumptions for K, X, Y, g, E, Λ as in (3.i) and the just mentioned particularization, if the sets $Q^*(t,\xi;z)$ are closed and convex, and have property (K) with respect to z in $\mathbb{R}^r$, then Φ is a weakly sequentially compact subset of K in X.

This remark is related to the reduction of property (Q) to property (K) in optimization theory (closure theorems) in connection with weak convergence, and when the relevant sets lie in finite dimensional spaces (Cf. Cesari and Suryanarayana [7c], [7d]). We mention here that Goodman [12] has shown the equivalence of property (Q) with properties known in Convex Analysis, and that Suryanarayana [21a] has proved that Minty's and Brezis's maximal monotonicity in any Hilbert space implies property (Q).

4. The linear abstract evolution equation.

Let G be a fixed bounded domain in R^n, let V be any real Banach space of functions on G, let V^* be the dual of V, let H be a Hilbert space such that $V \subset H \subset V^*$ with dense continuous imbeddings, let $\langle\ ,\ \rangle$ denote the pairing of V and V^*, and $\|\ \|$, $\|\ \|_*$, $|\ |$ the norms in V, V^*, H respectively.

Let $J = [0,T]$ be a given interval, and for given p, q > 1, $1/p + 1/q = 1$, let $X = L_p(J,V)$, $X^* = L_q(J,V^*)$ denote the spaces of functions on J with values in V and V^* respectively and integrable L_p- and L_q-norms. Then X and X^* are dual spaces with usual pairing $x,y = \int_0^T \langle x(t),y(t)\rangle dt$. Let p = q = 2.

Let $a(t;v,w)$ be a bilinear form on $V \times V$ such that $|a(t;v,w)| \le L\|v\|\|w\|$ and $a(t;v,v) \ge \alpha\|v\|^2$ for some constants L, $\alpha > 0$, all $v,w \in V$, and a.a. $t \in J$. Then a defines, for a.a. $t \in J$, a linear operator $A(t):V \to V^*$ by $a(t;v,w) = \langle A(t)v,w\rangle$, and then $\|A(t)\| \le L$, $\langle A(t)v,v\rangle \ge \alpha\|v\|^2$, that is, $A(t)$ is coercive. Also, a linear operator Λ: $X \to X^*$ is defined by taking $(\Lambda x)(t) = A(t)x(t)$. Then $\{\Lambda x,x\} \ge \alpha\|x\|_X^2$ for $x \in X$. It is assumed that Λ is weakly continuous in X, i.e., $x_k \rightharpoonup x$ implies $\Lambda x_k \rightharpoonup \Lambda x$ in the weak star topology of X^*.

Let W denote the space of all functions $x \in X = L_2(J,V)$ with distributional derivative $dx/dt \in X^* = L_2(J,V^*)$ and norm $\|x\|_X^2 = \|x\|_X^2 + \|dx/dt\|_{X^*}^2$. Then it is well known (see, e.g. [5]) that the operator E defined by $Ex = dx/dt$ with domain $D(E) = [x \in X,\ dx/dt \in X^*,\ x(0) = 0]$ is monotone and maximal monotone. In this situation, of course, $x(t)$, $t \in J$, or x: $J \to V$, is a continuous function of t in $J = [0,T]$, so that $x(0) = x(0+)$ is well defined (see, e.g. [19]). This has to be understood in the sense that x is actually an equivalent class, and this class contains one element, say y, which is continuous on J. We simply identify x with y.

With these assumptions and conventions, we conclude, as a corollary of (2.i), that the abstract Cauchy problem

(4.1) $$dx/dt + \Lambda x = f, \quad x(0) = 0,$$

$$x \in X = L_2(J,V),\ dx/dt \in X^* = L_2(J,V^*),\ t \in J = [0,T],\ f \in L_2(J,V^*),$$

has at least one weak solution $x \in X$, $\|x\| \le R = \alpha^{-1}\|f\|_{X^*}$.

Thereby, we have obtained the well known existence of at least one solution $z \in X$ to the Cauchy problem with initial data $x(0) = 0$ for the linear evolution equation, by sole topological considerations. Natural extensions make this existence statement essentially equivalent to others proved independently (see[5], and the end of no. 5 for some of the extensions mentioned in a more particular situation). An analytical existence proof is given, for instance, in Lions [18], where also uniqueness is proved.

As the reader may have noticed, the linearity of Λ plays here no role. We shall consider a quasilinear problem in no. 6.

5. The Cauchy-Dirichlet evolution equation in a cylinder with Λ a differential operator of order two.

As a way of an example, the following further particular case of nos. 2, 3, 4 is of interest. Let G be a bounded domain in the ξ-space $\mathbb{R}^n$, $\xi = (\xi_1,\cdots,\xi_n)$, let $J = [0,T]$ and let $a_{ij}(t,\xi)$, $i,j=1,\cdots,n$, be given bounded measurable functions on $J \times G$ such that

$$\Sigma_{i,j=1}^{n} a_{ij}(t,\xi)\zeta_i\zeta_j \geq \alpha(\zeta_1^2+\cdots+\zeta_n^2)$$

for all $(t,\xi) \in J \times G$, all $\zeta = (\zeta_1,\cdots,\zeta_n) \in \mathbb{R}^n$ and some constant $\alpha > 0$. Also, take $H = L_2(G,\mathbb{R})$ and let us take for V the Sobolev space $V = W_o^{1,2}(G)$, that is, briefly, the space of all functions $v(\xi)$, $\xi \in G$, whose traces on ∂G are zero, and v and the distributional derivatives $\nabla v = (\partial v/\partial\xi_1,\cdots,\partial v/\partial\xi_n)$ are all L_2-integrable functions on G. We can take in V the norm $\|v\| = \|v\|_V = \|\nabla v\|_{L_2(G)}$. Finally, let us take

$$a(t,v,w) = \int_G \Sigma_{i,j=1}^{n} a_{ij}(t,\xi)(\partial v/\partial\xi_j)(\partial w/\partial\xi_i), \quad v,w, \in V,$$

and $p = q = 2$, $X = L_2(J,V)$. Now the elements x of $X = L_2(J,V)$ are functions, say $x(t)$ on J with values in $V = W_o^{1,2}(G)$, or equivalently functions $x(t,\xi)$ on $J \times G$ with values in $\mathbb{R}$, with x and $\nabla x = (\partial x/\partial\xi_1,\cdots,\partial x/\partial\xi_n)$ both L_2-integrable in $J \times G$, and norm

$$\|x\|_X = \int_J \int_G |\nabla x|^2_{L_2(J\times G)} dt d\xi = \int_J \|x(t,\cdot)\|_V^2 \, dt.$$

Note that the linear operator Λ: $X \to X^*$ corresponding to the quadratic form a is now implicitely defined by the relation

$$\{\Lambda x,y\} = \int_J\int_G \Sigma_{i,j=1}^{n} a_{ij}(t,\xi)(\partial x/\partial\xi_j)(\partial y/\partial\xi_i)dt d\xi, \quad x,y \in X, \tag{5.1}$$

and Λ has the desired property

$$\{\Lambda x,x\} = \int_J\int_G \Sigma_{i,j=1}^{n} a_{ij}(t,\xi)(\partial x/\partial\xi_j)(\partial x/\partial\xi_i)dt d\xi$$
$$\geq \alpha \int_J\int_G |\nabla x|^2 dt d\xi = \alpha \int_J \|x(t,\cdot)\|_V^2 dt = \alpha\|x\|_X^2.$$

Moreover Λ is clearly weakly continuous as a map from X to X^*. Now the operator E: $D(E) \subset X \to X^*$ is the operator defined by $Ex = dx/dt$ with domain $D(E) = [x \in X,\ dx/dt \in X^*,\ x(0) = 0]$, or equivalently $Ex = \partial x/\partial t$, and for $x \in D(E)$ the distributional derivative $\partial x/\partial t$ is now a function in $L_2(J \times G,\mathbb{R})$.

Thus, all conditions of (2.i) are satisfied and the equation $dx/dt + \Lambda x = f$ has at least one weak solution $x \in X$ with $|x|_X \leq R$ and $R = \alpha^{-1}|f|_*$. By a weak solution x we mean, of course that

(5.2) $\int_J\int_G(\partial x/\partial t)w\,dtd\xi + \int_J\int_G\Sigma_{i,j=1}^n a_{ij}(t,\xi)(\partial x/\partial\xi_j)(\partial w/\partial\xi_i)dtd\xi = \int_J\int_G f(t,\xi)w\,dtd\xi$

for all $w \in X$, $w = w(t,\xi)$. Whenever x and the coefficients a_{ij} are sufficiently smooth, then since $x = 0$, $w = 0$ on ∂G, by integration by parts we have a solution x of the equation (in usual notations)

(5.3) $\partial x/\partial t - \Sigma_{i,j=1}^n(\partial/\partial\xi_i)(a_{ij}(t,\xi)(\partial x/\partial\xi_j)) = f(t,\xi)$, $(t,\xi) \in [0,T] \times G$,

$$x(0,\xi) = 0,\ \xi \in G; \quad x(t,\xi) = 0,\ (t,\xi) \in [0,T] \times \partial G.$$

The extension to the more general case $x(0,\xi) = x_o(\xi)$, $x_o \in V$, is immediately proved by taking $x = x_o + y$ and equation (4.2) is then replaced by $dy/dt + \Lambda(x_o + y) = f$, $y(0,\xi) = 0$. The further extension to the case $x_o \in L_2(G)$ and $f \in V^*$ can be handled, by a passage to the limit, by an argument which is similar to one in Lions [18] for an analogous situation (Cf. [5] for details).

The same equation (5.3) with terms of order one and zero, namely

$$\partial x\partial t - \Sigma_{i,j=1}^n(\partial/\partial\xi_i)(a_{ij}(t,\xi)\partial x/\partial\xi_j)) + \Sigma_{i=1}^n b_i(t,\xi)(\partial x/\partial\xi_i)$$
$$+ c(t,\xi)\,x(t,\xi) = f(t,\xi),$$

where all coefficients a_{ij}, b_i, c are bounded measurable functions on $J \times G$, and still $\Sigma_{ij}a_{ij}\zeta_i\zeta_j \geq \alpha(\zeta_1^2+\cdots+\zeta_n^2)$, can be handled analogously by well known algebraic manipulations connected with the Garding inequality (Cf. [18] and for details see also [5]).

An analogous treatment holds for the Cauchy-Neumann problem where it is required that the normal derivative $\partial x/\partial\nu$ on ∂G is zero. Here we take for V the corresponding subspace of the Sobolev space $W^{1,2}(G)$ whose normal derivative as an element of $W^{1/2,2}(\partial G)$ is zero (Cf. [5], where also extensions, analogous to the ones above, are considered).

6. <u>The Cauchy-Dirichlet problem for the quasi linear evolution equation.</u> Again, as a particular case, let $Y = L_1(J,L_1(G,\mathbb{R})) = L_1(J \times G,\mathbb{R})$, let the coefficients $a_{ij}(t,\xi,y)$ be real-valued Carathéodory functions in $J \times G$, that is, measurable in (t,ξ) for all $y \in \mathbb{R}$, and continuous in y for a.a. $(t,\xi) \in J \times G$. Let the same coefficients be bounded in $J \times G$, say $|a_{ij}(t,\xi,y)| \leq C$ for some constant C, and

(6.1) $$\Sigma_{ij}\, a_{ij}(t,\xi,y)\zeta_i\zeta_j \geq \alpha(\zeta_1^2+\cdots+\zeta_n^2)$$

for all $\zeta = (\zeta_1,\cdots,\zeta_n) \in \mathbb{R}^n$, all $y \in \mathbb{R}$, a.a. $(t,\xi) \in J \times G$, and some constant $\alpha > 0$.

For $V = W_o^{1,2}(G)$, $H = L_2(G)$, $X = L_2(J,V)$ the operator $\Lambda: X \to X^*$ implicitly defined by the relation

$$\{\Lambda x,y\} = \textstyle\int_J\int_G \Sigma_{i,j=1}^n\, a_{ij}(t,\xi,(Mx)(t,\xi))(\partial x/\partial\xi_j)(\partial y/\partial\xi_i), \quad x,\ y \in X,$$

certainly has the needed property $\{\Lambda x,x\} \geq \alpha\|x\|_X^2$ as in No. 5. It remains to prove that Λ is weakly continuous as a map from X to X^*. This is a consequence of the following lemma.

(6.i) (lemma) If S has finite measure in $\mathbb{R}^n$, if $f_k \to f$ weakly in $L_p(S,\mathbb{R})$, $p \geq 1$, if $|f_k(\xi)| \leq C$ for all $\xi \in S$, and $h_k \to h$ pointwise a.e. in S with $h_k, h \in L_\infty(S,\mathbb{R})$, then $f_k h_k \to fh$ weakly in $L_p(S,\mathbb{R})$. (Cf. Cesari [4e] II, p. 484).

Indeed, by the above considerations, if $x_k \to x$ weakly in $X = L_2(J,V)$, then $Mx_k \to Mx$ strongly in $L_1(J,Y)$, that is, $a_{ij}(t,\xi,(Mx_k)(t,\xi)) \to a_{ij}(t,\xi,(Mx)(t,\xi))$ in $L_1(J \times G,\mathbb{R})$ as well as pointwise a.e. in $J \times G$, with $|a_{ij}(t,\xi,(Mx_k)(t,\xi))| \leq C$, and by the lemma, with $S = J \times G$, n replaced by $n+1$, and ξ by (t,ξ), we derive that $\Lambda x_k \to \Lambda x$ weakly in L_2.

Let $U = \mathbb{R}^\mu$, $\mu \geq 1$, let $\omega_o(t,\xi)$ be a measurable multifunction $\omega_o: J \times G \to \mathbb{R}$, and let $u(t,\xi) \in \omega_o(t,\xi)$ denote any measurable selection of ω_o. Let $g(t,\xi,z,u)$, or $g: J \times G \times \mathbb{R} \times \mathbb{R}^\mu \to \mathbb{R}$ be a Carathéodory function, that is, measurable in (t,ξ) for all (z,u) and continuous in (z,u) for a.a. (t,ξ), and such that $|g(t,\xi,z,u)| \leq \ell(t)$, $t \in J$, for some fixed $\ell \in L_2(J,\mathbb{R})$. Then, for every measurable selection u of ω_o, there is a solution $x \in X$ to the equation (in weak form)

$$\int_J\int_G (dx/dt)y\,dt\,d\xi + \{\Lambda x,y\} = \int_J\int_G g(t,\xi,(Mx)(t,\xi),u(t,\xi))y\,dt\,d\xi,$$

$$x(0,\xi) = 0,\ \xi \in G,\quad x(t,\xi) = 0,\ (t,\xi) \in J \times \partial G,\quad x \in X,\ y \in X.$$

Thus, in usual notation, we conclude with the following statements

(6.ii) (Cesari and Hou [5]). Let the coefficients $a_{ij}(t,\xi,z)$ be measurable in (t,ξ) for every $z \in \mathbb{R}$, continuous in z for a.a. $(t,\xi) \in J \times G$, bounded in $J \times G$, and satisfy relation (6.1). Let $g(t,\xi,z,u)$ be measurable in (t,ξ) for every (z,u), continuous in (z,u) for a.a. $(t,\xi) \in J \times G$, and $|g(t,\xi,z,u)| \leq \ell(t)$ for some $\ell \in \ell_2(J,\mathbb{R})$. Let the map $M: X \to L_1(J \times G)$ transform weak convergent sequences in X into strong convergent sequences in $L_1(J \times G)$. Then, for every measurable selection u of ω_o, the Cauchy problem for the quasi linear evolution equation (in usual notations)

$$\partial x/\partial t - \Sigma_{ij}(\partial/\partial\xi_i)(a_{ij}(t,\xi,(Mx)(t,\xi))\partial x/\partial\xi_j) = g(t,\xi,(Mx)(t,\xi),u(t,\xi)),$$

$$(t,\xi) \in J \times G,$$

$x(0,\xi) = 0$, $\xi \in G$; $x(t,\xi) = 0, (t,\xi) \in J \times \partial G$, $x \in X$, has at least one solution x with $\|x\|_X \leq R = \alpha^{-1}\|\ell\|_2$.

We may now consider the class $\Phi = \{x\}$ of all solutions $x \in X$, $\|x\| \leq R$, for all possible selections u of ω_o. Let us assume, for instance, that ω_o is a fixed compact polyhedron in $\mathbb{R}$ (closed interval if $\mu = 1$). Here g is a scalar function, and because of the continuity of $g(t,\xi,x,u)$ in (z,u), the sets

$$Q^*(t,\xi,z) = [\zeta = g(t,\xi,z,u),\ u \in \omega_o] \subset \mathbb{R},$$

are closed segments, hence closed and convex, and certainly have property (K) with respect to z (Cf. Cesari [4abc]). Thus, for ω_o a fixed compact polyhedron in $\mathbb{R}$, and the same assumptions as in (6.ii), the class $\Phi = \{x\}$ is weakly sequentially compact in X.

Extensions as mentioned in no. 5 can be easily obtained, including the corresponding Cauchy-Neumann problem (Cf. [5]).

7. Existence of optimal solutions for control problems of the Mayer type.

We consider here the problem of the absolute minimum of a functional $I[z]$ in the class Φ, that is, the class of all elements $x(t)$, $t \in J$, $x \in K$, $x \in X$, which together with some measurable function $u(t)$, $t \in J$, $u(t) \in U$, satisfy

$$(7.1) \qquad Ex(t) + \Lambda x(t) = g(t,(Mx)(t),u(t)), \; u(t) \in \omega(t), \; t \in J \text{ (a.e.)}.$$

The following existence theorem is now an immediate consequence of (3.i):

(7.i) Theorem (an existence theorem for Mayer problems). Under hypotheses (A_1) - (A_5), if $I[x]$ is weakly lower semicontinuous in $X = L_p(J,V)$, and is bounded below on every bounded subset of X, then $I[x]$ has an absolute minimum in Φ.

For a variant of (7.i) we need alternate assumptions:

(C_1) Let U be a separable Banach space, and let $\omega: J \to U$ be a measurable multifunction with closed convex values, such that every $\omega(t)$ is a subset of a fixed closed convex subset A of U, and such that every bounded subset of A is relatively weakly compact in U.

(C_2) Every measurable selection u of ω satisfies $\|u\|_{L^r(J,U)} \le M$, $1 < r < +\infty$, for some constants M and r, and the set $G(t) = [(z,v) \mid z = g(t,z,v),\; v \in \omega(t)]$ is closed and convex for a.a. $t \in J$.

As usual we denote by Ω the class of all admissible pairs x,u.

(7.ii) Theorem (an existence theorem for Mayer problem) (S. H. Hou [14d]). Under hypotheses (A_1) - (A_5), (C_1), (C_2), if $I[x,u]$ is a weakly lower semicontinuous functional on $L_p(J,V) \times L_r(J,U)$, $1 < p,\ r < +\infty$, and $I[x,u]$ is bounded below on every bounded subset of $L_p(J,V) \times L_r(J,U)$, then I has an absolute minimum in Ω.

8. Existence of optimal solutions for control problems of the Lagrange type.

We are concerned here with the minimum of the functional

$$I[x,u] = \int_J f_o(t, Mx(t), u(t))dt$$

in the class $\Omega = \{(x,u)\}$ of all admissible pairs, i.e., such that

$$Ex(t) + \Lambda x(t) = g(t,(Mx)(t),u(t)), \; u(t) \in \omega(t) \subset U, \; x \in K \subset X = L_p(J,V)$$

Here we need the following assumptions

(L_1) Let Y be a real Banach space, and f_o a Carathéodory map from $J \times Y \times U \to \mathbb{R}$, i.e., $f_o(t,y,u)$ is strongly measurable in t for every (y,u) and is continuous in (y,u) for almost all $t \in J$.

(L_2) There are $b \in V$ and $\psi \in L_1(J,\mathbb{R})$, $\psi(t) \geq 0$, such that $f_o(t,y,u) \geq \langle b, g(t,u)\rangle - \psi(t)$ for all $(t,y,u) \in J \times Y \times U$.

(L_3) The multifunction $\tilde{Q}: J \times Y \to \mathbb{R} \times V^*$ defined by $\tilde{Q}(t,y) = \{(z^o,z) \mid z^o \geq f_o(t,y,u), \; z = g(t,y,u), \; u \in \omega(t) \text{ a.e.}\}$ has property (Q) with respect to y for almost all $t \in J$. Thus, the sets $Q(t,y)$ are convex in $\mathbb{R} \times V^*$.

(L_4) $J[x,u] < +\infty$ for some $(x,u) \in \Omega$.

(8.i) Theorem (an existence theorem for Lagrange problems) (S. H. Hou [14d] and L. Cesari [leII]). Under hypotheses $(A_1) - (A_5)$, $(L_1) - (L_5)$, the functional $I[x,u]$ has an absolute minimum in Ω.

As at the end of no. 3, when V and V^* are spaces of finite dimensional vector valued functions on a finite dimensional domain G in $\mathbb{R}^n$, then the contentions of theorems (7.i) and (8.i) hold under essentially weaker assumptions. Namely, we may require only property (K) instead of (Y) for suitable subsets Q^* of finite dimensional spaces.

Thus, let G be a domain in $\mathbb{R}^n$, let us consider (3.1) as a problem in the cylinder $J \times G$, and let us make the same assumptions at the end of no. 3. Now g is a function with values in $\mathbb{R}^r$ of (t,ξ,z,u) in $J \times G \times \mathbb{R}^s \times \mathbb{R}^\mu$. We shall need here the sets $Q^*(t,\xi,z)$ of $\mathbb{R}^r$ defined by

$$Q^*(t,\xi,z) = [\zeta \in \mathbb{R}^r \mid \zeta = g(t,\xi,z,u), u \in \omega_o(t,\xi)].$$

Then (7.i) holds under requirement analogous to $(A_1 - A_4, A_5')$. Correspondingly, in no. 8, f_o is a function with values in $\mathbb{R}$ of (t,ξ,z,u) in $J \times G \times \mathbb{R}^s \times \mathbb{R}^\mu$.

We shall need the sets $\tilde{Q}^*(t,\xi,z)$ of $\mathbb{R}^{r+1}$ defined by

$$\tilde{Q}^*(t,\xi,x) = [(\zeta^o,\zeta) \in \mathbb{R}^{r+1} \mid \zeta^o \geq f_o(t,\xi,z,u),\ \zeta = g(t,\xi,z,u), u \in \omega_o(t,\xi) \subset \mathbb{R}^{\mu}].$$

Then (8.i) holds under requirements analogous to (A_1-A_4,A_5', L_1-L_3) when in (A_5'), in particular, we require that for a.a. $(t,\xi) \in J \times G$, the sets $\tilde{Q}^*(t,\xi,z)$ are closed and convex and have property (K) with respect to z in $\mathbb{R}^{r+1}$.

9. Existence of solutions as a direct consequence of fixed point theorems.

We shall now use directly the fixed point theorem for multifunctions to obtain existence theorems for solutions of functional control equations of the form

$$(9.1) \qquad \begin{aligned} &Ex(t) = g(t,Mx(t),u(t)), \quad t \in [0,T] \text{ (a.e.)}, \\ &u(t) \in \omega(t), \quad t \in [0,T] \text{ (a.e.)}, \end{aligned}$$

where u is a measurable selection of ω, and x belongs to a given space X. A pairs x, u satisfying (9.1) is called admissible, and then x will be called an admissible trajectory or simply a trajectory.

Of course the operator E above may well be defined in a certain domain D(E) of functions x(t), $0 \leq t \leq T$, with values in a given space V satisfying certain properties. In addition other requirements can be imposed on the solutions x(t) of (9.1). In other words, a nonempty class ψ of elements x(t), $0 \leq t \leq T$, $x \in X$, may be stipulated as the class of the desired solutions, or acceptable solutions. Then, we shall say that (9.1) is <u>ψ-controllable</u> if there is some admissible pair x, u with $x \in \psi$.

Let $J = [0,T]$, let V be a separable reflexive Banach space, let V^* be the dual of V, and let $X = L_p(J,V)$, $X^* = L_q(J,V^*)$, $1 < p < +\infty$, $1/p + 1/q = 1$. Let $\|\ \|_V, \|\ \|_{V^*}, \|\ \|_X$ denote the norms in V, V^*, X respectively. We shall now use the same notations as in no. 3, in particular the sets Q(t,z) defined there.

Let K be a nonempty subset of ψ, thus $K \subset \psi \subset D(E) \subset X$, and let $\Gamma: K \to V$, or $x \to \Gamma(x)$, be the multifunction defined by

$$\Gamma(x) = [z \in K \mid Ez(t) \in Q(t,Mx(t)),\ t \in [0,T] \text{ (a.e.)}]$$

If Γ has a fixed point in K, or $x \in \Gamma(x) \subset K$, then $Ex(t) \in Q(t,Mx(t))$, $t \in J$, and, as we will see, $x(t)$, $0 \leq t \leq T$, is a solution of (9.1).

We shall need now the specific hypotheses:

(H_1) K is a nonempty closed convex bounded subset of ψ, i.e. $K \subset \psi \subset D(E) \subset X$.

(H_2) The sets $Q(t,z)$ are closed and convex, and for a.a. $t \in J$, have property (Q) with respect to z in Z.

(H_3) For every bounded sequence $[b_k]$ in $L_p(J,Y)$ and for every bounded sequence $[v_k]$ in X with $Ev_k(t) \in Q(t,b_k(t))$, the sequence $[Ev_k]$ is bounded in X^*.

Certainly (H_3) is satisfied if $|g(t,z,u)|_{Y^*} \leq \beta(t) + \lambda|z|_Y^{p/q}$ for $\beta \in L_q(J,\mathbb{R})$. Indeed, by work of Nemitsky, Krasnoselskii et al., this is the necessary and sufficient condition in order that bounded sequences are mapped into bounded sequences.

Note that under these assumptions, the multifunction Γ: $K \to K$ has convex values. Indeed, if z_1, $z_2 \in \Gamma(x)$ for some $x \in K$, let $z = \lambda z_1 + (1-\lambda)z_2$ and note that $Ez_i(t) = g(t,Mx(t), u_i(t))$, $i = 1,2$, $t \in J$, for some measurable selections $u_1(t), u_2(t)$ of $\omega(t,Mx(t))$. Thus

$$\begin{aligned} Ez(t) &= \lambda Ez_1(t) + (1-\lambda)Ez_2(t) \\ &= \lambda f(t,Mx(t),u_1(t)) + (1-\lambda)f(t,Mx(t),u_2(t)). \end{aligned}$$

Since both $f(t,Mx(t),u_1(t))$ and $f(t,Mx(t),u_2(t))$ belong to the convex set $Q(t,Mx(t))$, then $Ez(t) \in Q(t,Mx(t))$, $t \in J$, and $z \in \Gamma(x)$ by the implicit function theorem, as formulated by Hou in [14b], there is a measurable selection u of ω such that $Ez(t) = g(t,(Mx)(t),u(t))$, $t \in J$ (a.e.).

Under the above assumptions the multifunction Γ: $K \to K$ has a closed graph in $X_w \times X_w$, and then the following theorem can be derived from (1.i):

(9.i) <u>Theorem</u> (A sufficient condition for ψ-controllability)(S. H. Hou [14c]). For $K \subset \psi \subset X$ and $\Gamma(z) \neq \emptyset$ for every $z \in K$, then system (9.1) is ψ-controllable.

Under the conditions described at the end of no. 3 requirement (H_2) concerning property (Q) for the sets Q, can be replaced by the weaker requirement (H_2') that for almost all $(t,\xi) \in J \times G$ the sets $Q^*(t,\xi,z)$ have property (K) with respect to z in $\mathbb{R}^s$.

Remark 1. For subsets Q(t,x) in finite dimensional spaces, closure theorems in Optimization Theory hold under sole property (K) because of the following remark. First, if $\tau_k(t)$, $t \in \subset R^n$, $k = 1,2,\cdots$, $\tau_k \in (L_1(S))^r$, and $\tau_k \rightharpoonup \tau$ weakly in $(L_1(S))^r$, then there is some scalar convex monotone positive function $\phi(\xi)$, $0 \leq \xi < +\infty$, with $\phi(\xi)/\xi \to +\infty$ as $\xi \to +\infty$, and $\lambda_k = \phi(|\tau_k(t)|)$ converges weakly to some function λ in $L_1(S)$ (De La Vallée Poussin). Now assume that $\tau_k(t) \in Q(t,x_k(t)) \subset \mathbb{R}^r$, $t \in J$, $k = 1,2,\cdots$, where $x_k \to x$ strongly in $(L_1(G))^s$, and where the sets Q(t,z) are closed and convex, and have property (K) with respect to z. Then the augmented sets in $\mathbb{R}^{r+1}$ defined by $R(t,z) = [(\xi^o,\xi) \mid \xi^o \geq \phi(|z|),\ \xi \in Q(t,z)]$ are convex and have property (Q) with respect to z (Cesari [4gh]). On the other hand $(\lambda_k(t), \tau_k(t)) \in R(t,x_k(t))$, $t \in S$, $k = 1,2,\cdots$, with $(\lambda_k,\tau_k) \rightharpoonup (\lambda,\tau)$ weakly. From Cesari's closure theorems [4d] then $(\lambda(t),\tau(t)) \in R(t,x(t))$ and $\tau(t) \in Q(t,x(t))$ a.e. in J. (Cf. Cesari, [4h]).

Remark 2. The use of fixed point theorems in questions of controllability for ordinary differential equations was initiated by Tarnove [22], and later continued by Dauer [8] and Henry [13]. For questions of controllability for functional differential equations Angell [1] made systematic use of fixed point theorems, as well as of properties (Q) and (K). (Cf. Angell's paper also for the bibliography). Finally, Hou's statement (9.i) concerns operator equations, and therefore may apply to partial differential equations.

References

[1] T. S. Angell, On controllability for nonlinear hereditary systems: a fixed point approach. Nonlinear Analysis 4, 1980, 529-546.

[2] H. Brezis, (a) Equations et inequations nonlineaires dans les espaces vectoriels en dualité, Ann. Inst. Fourier 18, 1968, 115-175. - (b) On some degenerate nonlinear parabolic equations. Nonlinear functional analysis. Proc. Symp. Pure Math., Amer. Math. Soc., 18, 1970, 28-38. - (c) Operateurs maximaux monotones, Amer. Elsevier, New York 1973.

[3] H. F. Bohnenblust and S. Karlin, On a theorem of Ville. Contributions to the Theory of Games (Kuhn and Tucker, eds.), Princeton Univ. Press 1950, vol. 1, pp. 155-160.

[4] L. Cesari, (a) Existence theorems for weak and usual solutions in Lagrange problems with unilateral constraints. Trans. Amer. Math. Soc. 124, 1966, 369-412, 413-429. - (b) Seminormality and upper semicontinuity in optimal control, Journ. Optimization Theory Appl. 6, 1970, 114-137. - (c) Closure, lower closure, and semicontinuity theorems in optimal control, SIAM J. Control 9, 1971, 287-315. - (d) Closure theorems for orientor fields and weak convergence. Archive Rat. Mech. Anal. 55, 1974, 332-356. - (e) Geometric and analytic views in existence theorems for optimal control in Banach spaces. I,II,III. Journ. Optimization Theory Appl. 14, 1974, 505-520; 15, 1975, 467-497; 19, 1976, 185-214. - (f) Functional analysis, nonlinear differential equations and the alternative method. Nonlinear Functional Analysis and Differential Equations. (L. Cesari, R. Kannan, J. D. Schuur, eds.), M. Dekker, New York 1976, pp. 1-197. - (g) Existence theorems for optimal controls of the Mayer type. SIAM J. Control 6, 1968, 517-552. - (h) Optimization Theory and Applications: Problems with Ordinary Differential Equations. Springer Verlag 1982.

[5] L. Cesari and S. H. Hou, Existence of solutions and existence of optimal solutions. To appear.

[6] L. Cesari and R. Kannan, (a) Functional analysis and nonlinear differential equations, Bull. Am. Math. Soc. 79, 1973, 1216-1219. - (b) An abstract existence theorem at resonance, Proc. Amer. Math. Soc. 63, 1977, 221-225. - (c) An existence theorem for periodic solutions of nonlinear parabolic equations. Rend. Accad. Milano. To appear. - (d) Periodic solutions of nonlinear wave equations with damping. Rend. Circolo Mat. Palermo. To appear. - (e) Periodic solutions of nonlinear wave equations with nonlinear damping. Communications in partial differential equations. To appear. - (f) Existence of solutions of nonlinear hyperbolic equations. Annali Scuola Norm. Sup. Pisa (4)6, 1979, 573-592; 7, 1980, p. 715. - (g) Solutions of nonlinear hyperbolic equations at resonance. Nonlinear Analysis.

[7] L. Cesari and M. B. Suryanarayana (a) Existence theorems for Pareto optimization. Multivalued and Banach space valued functions. Trans. Amer. Math. Soc. 244, 1978, 37-65. - (b) An existence theorem for Pareto problems. Nonlinear Analysis, 2, 1978, 225-233. - (c) Upper semicontinuity properties of set valued functions. Nonlinear Analysis, 4, 1980, 639-656. - (d) On recent existence theorems in the theory of optimization. Journ. Optimization Theory Appl. 31, 1980, 397-415.

[8] J. P. Dauer, A controllability technique for nonlinear systems, J. Math. Analysis Appl. 37, 1972, 442-451.

[9] N. Dunford and J. T. Schwartz, Linear Operators I, Interscience 1958.

[10] K. Fan, Fixed point and minimax theorems in locally convex topological linear spaces. Proc. Nat. Acad. Sci. U.S.A. 58, 1952, 121-126.

[11] I. Glicksberg, A further generalization of Kakutani fixed point theorem with application to Nash equilibrium points. Proc. Amer. Math. Soc. 3, 1952, 170-174.

[12] G. S. Goodman, The duality of convex functions and Cesari's property (Q). J. Optimization Theory Appl. 19, 1976, 17-23.

[13] J. Henry, Étude de la controllabilité de certaines equations paraboliques non-lineaires, Proc. IFIP Working Conference on Distributed Parameter Systems, Rome 1976.

[14] S. H. Hou (a) On property (Q) and other semicontinuity properties of multi-functions. Pacific J. Math. To appear. - (b) Implicit functions theorem in topological spaces, Applicable Mathematics. To appear. - (c) Controllability and feedback system. To appear. - (d) Existence theorems of optimal control problems in Banach spaces. Nonlinear Analysis. To appear.

[15] S. Kakutani, A generalization of Brouwer's fixed point theorem, Duke Math. Journ. 8, 1941, 457-459.

[16] R. Kannan, (a) Une version stochastique du théoréme de Kakutani, C. R. Acad. Sc. Paris, t. 287, Ser. A, 551-552; - (b) Random correspondences and nonlinear equations. J. Multivariate Analysis 11, 1981, 230-243.

[17] N. Kenmochi, Existence theorems for certain nonlinear equations, Hiroshima Math. J. 1, 1971, 435-443.

[18] J. L. Lions, Optimal Control of Systems Governed by Partial Differential Equations, Springer Verlag 1971, xi+396.

[19] J. L. Lions and E. Magenes, Nonhomogeneous Boundary Value Problems and Applications, I,II,III, Springer Verlag 1972-73.

[20] G. J. Minty, (a) On the maximal domain of a monotone function, Mich. Math. J. 8, 1961, 135-137. - (b) Monotone nonlinear operators in Hilbert space, Duke Math. J. 29, 1962, 341-346. - On a monotonicity method for the solution of nonlinear equations in Banach spaces, Proc. Nat. Acad. Sci. U.S.A. 50, 1963, 1038-1041.

[21] M. B. Suryanarayana, (a) Monotonicity and upper semicontinuity, Bull. Amer. Math. Soc. 82, 1976, 936-938; (b) Remarks on existence theorems for Pareto optimality, Dynamical Systems, a Univ. of Florida Intern. Symposium (Bednarek and Cesari, eds.), Academic Press 1977, pp. 335-347.

[22] I. Tarnove, A controllability problem of nonlinear systems. Math. Theory of Control (Balakrishnan and Neustadt, eds.), Academic Press 1967, pp. 170-179.

DUAL VARIATIONAL METHODS
IN NON-CONVEX OPTIMIZATION AND DIFFERENTIAL EQUATIONS

I. EKELAND

Department of Mathematics, U.B.C.
and
CEREMADE, Université Paris-Dauphine

1. Introduction

One of the basic tools of convex analysis is duality theory. This is an automatic procedure for transforming an optimization problem (the primal) into another one (the dual), in the hope that the second will prove more tractable than the first.

The success of these methods in convex optimization (see [19] for general reference) has led to numerous investigations into the non-convex case. The problem is not so much to build a non-convex duality theory as to find suitable applications for it. In this respect, let us single out the work of Ekeland [10], Toland [20], [21], [22], Aubin-Ekeland [1], and Auchmuty [2].

In 1978, Clarke [4] introduced into a special problem of Hamiltonian mechanics a method of that kind, which was immediately extended to more general situations by Clarke-Ekeland [7]. The analogue for the wave equation was given in Brezis-Coron-Nirenberg [3], and an abstract formulation was described in Ekeland-Lasry [16].

In this lecture, we shall try to review the abstract formulation and the more recent developments. We shall concentrate on the main ideas, and we shall refer to the original papers for detailed proofs.

2. A duality theorem

Let V be a reflexive Banach space. A map $A \in \mathcal{L}(V,V^*)$ will be called <u>symmetric</u> iff $A^* = A$. If Q is a continuous quadratic form on V, it can be written in a unique way as:

(1) $$Q(v) = \frac{1}{2} \langle Av,v \rangle \qquad \text{with } A^* = A .$$

We are given a function $I : V \to \mathbb{R} \cup \{+\infty\}$ by

(2) $$I(v) = Q(v) + F(v)$$

where Q is a continuous quadratic form and F a lower semi-continuous convex function.

Note that I is <u>not</u> a convex function unless Q is positive, which we do not assume. Indeed, all the applications we will discuss relate to the case when Q has positive and negative subspaces.

From now on we will use representation (1) for Q, and we recall that the Fenchel conjugate of F is the convex l.s.c. function $F^* : V^* \to \mathbb{R} \cup\{+\infty\}$ defined by:

(3) $$F^*(v^*) = \text{Sup } \{\langle v,v^* \rangle - F(v) \mid v \in V\}$$

If F is differentiable, and the supremum on the right-hand side is attained, we get the usual formula for the Legendre transform:

(4) $$F^*(v^*) = \{\langle v,v^* \rangle - F(v) \mid v^* = F'(v)\}$$

In the general, non-differentiable case, we shall denote by $\partial F(v)$ the subgradient of F at v. We recall the reciprocity formula:

(5) $$\begin{aligned} v^* \in \partial F(v) &\iff F(v) + F^*(v^*) - \langle v,v^* \rangle \leq 0 \\ &\iff v \in \partial F^*(v^*) \end{aligned}$$

We shall say that $\bar{v}$ is a <u>critical point</u> of I iff $A\bar{v} + \partial F(\bar{v}) \ni 0$.

The value $I(\bar{v}) = Q(\bar{v}) + F(\bar{v})$ will be called a <u>critical value</u> of I.

<u>Proposition 1</u> Consider the two functions I and J on V:

(6) $$I(u) = \frac{1}{2} \langle Au,u \rangle + F(u)$$

(7) $$J(v) = \frac{1}{2} \langle Av,v \rangle + F^*(-Av)$$

If $\bar{u}$ is a critical point of I, then it is also a critical point of J. If $\bar{v}$ is a critical point of J, and if the interior of $A(V) + \mathrm{dom}F^*$ contains the origin in V^*, then there is some $w \in \mathrm{Ker}\, A$ such that $\bar{v} - \bar{w} = \bar{u}$ is a critical point of J. Moreover

(8) $$J(\bar{v}) = -I(\bar{u}) \quad \square$$

<u>Proof</u> Assume $\bar{u}$ is a critical point of I:

$$A\bar{u} + \partial F(\bar{u}) \ni 0$$

This can also be written:

$$-A\bar{u} \in \partial F(\bar{u})$$

By the reciprocity formula:

$$\bar{u} \in \partial F^*(-A\bar{u})$$

Applying A to both sides:

$$A\bar{u} \in A\partial F^*(-A\bar{u}) .$$

Now it is known that $-A\partial F^*(-A\bar{u}) \subset \partial G(\bar{u})$, with $G(u) = F^*(-Au)$. The converse inclusion will hold if a technical condition is satisfied, namely that $A(V) + \mathrm{dom}\, F^*$ contains 0 in its interior. But here the first inclusion is enough:

$$A\bar{u} + \partial G(\bar{u}) \ni 0 ,$$

so $\bar{u}$ is a critical point of J.

Conversely, if $\bar{v}$ is a critical point of J, and the technical condition is satisfied, we can write

$$A\bar{v} - A\partial F^*(-A\bar{v}) \ni 0$$

This means that there exists some $\bar{u} \in \partial F^*(-A\bar{v})$ such that $A(\bar{v} - \bar{u}) = 0$. Setting $\bar{v} - \bar{w} = \bar{u} \in \operatorname{Ker} A$ and applying the reciprocity formula once more, we get:

$$-A\bar{v} \in \partial F(\bar{u})$$

But $A\bar{v} = A\bar{u}$, so we finally get:

$$A\bar{u} + \partial F(\bar{u}) \ni 0$$

All that remains to be shown is the statement about the critical values. Since $-A\bar{v} \in \partial F(\bar{u})$, we have:

$$F^*(-A\bar{v}) + F(\bar{u}) + \langle A\bar{v}, \bar{u}\rangle = 0 .$$

Since A is symmetric and $A\bar{v} = A\bar{u}$, this can be rewritten as:

$$F^*(-A\bar{v}) + \frac{1}{2}\langle A\bar{v}, \bar{v}\rangle + F(\bar{u}) + \frac{1}{2}\langle A\bar{u}, \bar{u}\rangle = 0$$

We recognize this as $J(\bar{v}) + I(\bar{u}) = 0$ □

It follows that I has critical points if and only if J has critical points. It is an essential feature of this transformation, though, that it <u>changes the type</u> of critical points: a saddle-point of I may very well correspond to a minimum of J. Let us illustrate this in finite dimension.

Let $V = \mathbb{R}^n$, and assume A is nondegenerate. Recall that the <u>index</u> of the quadratic form Q is the number of negative squares in any diagonalization of Q, and that the index of a critical point for a C^2 function is the index of the Hessian at that point. Local minima, for instance, have index 0, and nondegenerate local maxima have index n.

<u>Lemma 2</u> Let k be the index of Q. Let $\bar{u}$ and $\bar{v}$ be corresponding

critical points of I and J. Assume that F and F^* are C^2 near $\bar{u}$ and $\bar{v}$, and that $F''(\bar{u})$ is positive definite. If either $\bar{u}$ or $\bar{v}$ is a local minimum, the other one is a saddle-point with index k. If either $\bar{u}$ or $\bar{v}$ is a local maximum, the other one is a saddle-point with index $n-k$. □

<u>Proof</u> In this situation, we can replace I and J by their Hessian at $\bar{u}$ and $\bar{v}$, namely:

$$\langle I''(\bar{u})u,u\rangle = \langle Au,u\rangle + \langle Bu,u\rangle$$

$$\langle J''(\bar{v})v,v\rangle = \langle Av,v\rangle + \langle B^{-1}Av,Av\rangle$$

with $B = B^* = F''(\bar{u})$ positive definite.

We diagonalize all these quadratic forms together:

$$\langle I''(\bar{u})u,u\rangle = \sum_{i=1}^{n} a_i u_i^2 + \sum_{i=1}^{n} b_i u_i^2$$

$$= \sum_{i=1}^{n} (a_i + b_i) u_i^2$$

$$\langle J''(\bar{v})v,v\rangle = \sum_{i=1}^{n} a_i v_i^2 + \sum_{i=1}^{n} \frac{a_i^2}{b_i} v_i^2$$

$$= \sum_{i=1}^{n} \frac{a_i}{b_i} (a_i + b_i)\, v_i^2 .$$

Since $b_i > 0$ for all i, and k of the a_i are negative, it is clear that precisely k of the squares change sign. The result follows immediately. □

This opens the possibility that a saddle-point of I (or J) is transformed into a minimum of J (or I), and so becomes considerably easier to find.

We now proceed to illustrate this method in an infinite-dimensional setting. This may require some adjustments of proposition 1 (see [16],[17]), because the technical condition may be hard to satisfy in function spaces, but the general idea is the same.

Let us note (a remark due to Lasry) that the range of applications of proposition 1 will include some cases when F is not convex. Indeed, it is enough that $F(u) + k||u||^2/2$ be convex for suitably large $k > 0$.

<u>Proposition 3</u> Assume V is Hilbertian, $F : V \to \mathbb{R} \cup \{+\infty\}$ is l.s.c. and $\tilde{F}(u) = F(u) + k||u||^2/2$ is convex, with 0 in the interior of $((A - kI)V + \text{dom}\, \tilde{F}^*)$. Define

$$I(u) = \frac{1}{2} \langle Au,u\rangle + F(u)$$

$$J(v) = \frac{1}{2} \langle (A - kI)v,v\rangle + \tilde{F}^*(kv - Av)$$

Then I has critical points iff J has critical points. □

<u>Proof</u> Just write:

$$I(u) = \frac{1}{2} \langle Au,u\rangle - \frac{k}{2} \langle u,u\rangle + \tilde{F}(u)$$

$$= \frac{1}{2} \langle (A - kI)u,u\rangle + \tilde{F}(u)$$

and apply proposition 1. □

The same trick will give:

<u>Proposition 4</u> Assume V is Hilbertian, $G : V \to \mathbb{R} \cup \{+\infty\}$ is l.s.c. and $\tilde{G}(v) = G(v) + k||v||^2/2$ is convex. Assume $(I - kA) = A_k$ is invertible, and 0 belongs to the interior of $(AA_k(V) + \text{dom}\, \tilde{G})$. Define:

$$H(u) = \frac{1}{2} \langle AA_k^{-1}w,w\rangle + \tilde{G}^*(w)$$

$$K(v) = \frac{1}{2} \langle Av,v\rangle + G(-Av)$$

Then H has critical points iff K has critical points. □

<u>Proof</u> Rewrite the second function:

$$K(v) = \frac{1}{2} <Av,v> - \frac{k}{2} <Av,Av> + \tilde{G}(-Av)$$

$$= \frac{1}{2} <AA_k v,v> + \tilde{G}oA_k^{-1}(-AA_k v)$$

By proposition 1, this corresponds to:

$$\frac{1}{2} <AA_k u,u> + (\tilde{G}oA_k^{-1})*(u)$$

But $(\tilde{G}oA_k^{-1})* = \tilde{G}*oA_k$. Changing the variable to $w = A_k u$, we get the function H . □

3. Applications

A - Hamiltonian systems

There is by now a number of results by this method (see [5],[6],[8], [9],[11],[12],[13],[14],[15],[18]). We restrict ourselves to a typical situation (see [6],[14]).

Recall that a matrix $M \in L(\mathbb{R}^{2n})$ is <u>symplectic</u> if $M*JM = J$, where J is given by:

$$J = \begin{pmatrix} 0 & I_n \\ -I_n & 0 \end{pmatrix}$$

We are given a convex continuous function H on $\mathbb{R} \times \mathbb{R}^{2n}$ a symplectic matrix $M \in L(\mathbb{R}^{2n})$, and we consider the boundary-value problem:

$$(H_M) \qquad \dot{u}(t) = J\partial H(t,u(t)) , \qquad u(T) = Mu(0)$$

Write $\mathbb{R}^{2n}$ as $\mathbb{R}^n \times \mathbb{R}^n$, and u as (x,p) . When H is C^1 , we recognize the canonical form of Hamilton's equations:

$$\dot{x} = \frac{\partial H}{\partial y}(t,x,y) , \qquad \dot{y} = - \frac{\partial H}{\partial x}(t,x,p)$$

It is a well-known fact (the least action principle) that the solutions of (H_M) are precisely the critical points of the functional

$$\int_0^T [\tfrac{1}{2}(J\dot{u},u) + H(t,u)]dt = I(u)$$

on the subspace V of $H^1(0,T;\mathbb{R}^{2n})$ consisting of all functions $u(t)$ such that $u(T) = Mu(0)$.

<u>Proposition 5</u> Assume M is symplectic, and let $\bar{v}$ be a critical point of the functional

$$\int_0^T [\tfrac{1}{2}(J\dot{v},v) + H^*(t,-J\dot{v})]dt = J(v)$$

on the space

$$V = \{v \in H^1(0,T;\mathbb{R}^{2n}) \mid v(T) = Mv(0)\} .$$

Then there is some vector $\xi \in \mathrm{Ker}(M-I)$ such that $\bar{v}(t) + \xi = \bar{u}(t)$ is a solution of (H_M) □

The quadratic form on V is here:

$$Q(u) = \frac{1}{2}\int_0^T (J\dot{u},u)dt$$

To find its symmetric representation A, we integrate by parts:

$$\begin{aligned}
\langle Au_1,u_2\rangle &= \frac{1}{2}\int_0^T (J\dot{u}_1,u_2)dt + \frac{1}{2}\int_0^T (J\dot{u}_2,u_1)dt \\
&= \int_0^T (J\dot{u}_1,u_2)dt + (Ju_2(t),u_1(t))\Big|_0^T \\
&= \int_0^T (J\dot{u}_1,u_2)dt + (JMu_2(0),Mu_1(0)) - (Ju_2(0),u_1(0)) \\
&= \int_0^T (J\dot{u}_1,u_2)dt + (u_1(0),(M^*JM-I)u_2(0)) \\
&= \int_0^T (J\dot{u}_1,u_2)dt .
\end{aligned}$$

So $A = J\frac{d}{dt}$. The result then follows from proposition 1, suitably adapted to meet the technical condition.

Note two particularly interesting cases for (H_M):

$M = I$ (periodic solutions)

$M = -I$ (antiperiodic solutions)

B - Systems with gyroscopic forces or viscous friction

The previous analysis can easily be extended to more realistic mechanical systems, including gyroscopic forces or viscous friction.

Systems with gyroscopic forces are described by the following second-order equation

$$\ddot{x} + 2K\dot{x} + V'(t,x) = 0 \qquad \text{with } K^* = -K \in \mathcal{L}(\mathbb{R}^n)$$

Introducing the n-dimensional variable $y = \dot{x} + Kx$, and the Hamiltonian:

$$H(x,y) = \frac{1}{2} y^2 + V(t,x) + \frac{1}{2} \langle Kx,Kx \rangle + \langle Ky,x \rangle$$

we reduce this equation to the canonical system $\dot{u} = JH'(t,u)$.

Note that this Hamiltonian differs from the potential V by quadratic terms only. Set

$$\tilde{V}(t,x) = V(t,x) + \frac{1}{2} \langle Kx,Kx \rangle$$

<u>Proposition 6</u> Let K be antisymmetric and V convex. Let $\bar{v}$ be a critical point of the functional:

$$J(v) = \int_0^T [\frac{1}{2}(J\dot{v} + JKv,v) + \frac{1}{2} |\dot{x} + Kx|^2 + V^*(t,-\dot{y} - Ky)]dt$$

on the space:

$$V = \{v = (x,y) \in H^1(0,T;\mathbb{R}^{2n}) \mid u(0) = u(T)\}$$

Then there is some T-periodic solution $u_0 = (x_0, y_0)$ of the equation $\dot{u}_0 + Ku_0 = 0$ such that $x = \bar{x} + x_0$ solves the problem:

$$\begin{cases} \ddot{x} + 2K\dot{x} + V'(t,x) = 0 \\ x(0) = x(T) \quad \text{and} \quad \dot{x}(0) = \dot{x}(T) \end{cases} \qquad \square$$

Certain cases of viscous friction also fall within the scope of proposition 1. Consider the equation:

$$\ddot{x} + a\dot{x} + V'(t,x) = 0$$

We can find x and $\dot{x} = y$ as extremals of the integral

$$\int (-y\dot{x} + \frac{1}{2}y^2 + V(t,x))e^{at}dt$$

We refer to [17] for the appropriate duality procedure, using proposition 1. It turns out that the meaningful problem to solve is the following:

$$\begin{cases} \ddot{x} + a\dot{x} + V'(t,x) = 0 \\ x(0) = x(T) \quad \text{and} \quad \dot{x}(0) = e^{aT}\dot{x}(T) \end{cases}$$

C - A fourth-order problem

Consider the boundary-value problem (where ' denotes $\frac{d}{dt}$):

$$\begin{cases} (px'')'' - V'(t,x) = 0 \\ x^{(n)}(0) = x^{(n)}(T) \quad \text{for} \quad n = 1,2 \\ (px'')^{(n)}(0) = (px'')^{(n)}(T) \quad \text{for} \quad n = 3,4 \end{cases}$$

The solutions are the critical points of the functional

$$I(x) = \int_0^T [-\frac{1}{2} p|x''|^2 + V(t,x)]dt$$

on the space

$$V = \{x \in L^2(0,T;\mathbb{R}^n) \mid x'' \in L^2, x^{(n)}(0) = x^{(n)}(T),\ n = 1,2\}$$

We now have to wander a bit farther afield. The general theory of [16] (see [23] for details) gives the dual functional

$$J(y) = \int_0^T [-\frac{1}{2}(Sy,y) + V^*(t,y)]dt$$

on the space $L^2(0,T;\mathbb{R}^n)$, the compact linear operator $S: y \to u$ being defined by

$$\begin{cases} (pu'')'' = y \in L^2 \\ u^{(n)}(0) = u^{(n)}(T) & n = 1,2 \\ (pu)^{(n)}(0) = (pu)^{(n)}(T) & n = 3,4 \end{cases}$$

All this under the assumption that $V(t,\cdot)$ is convex and $0 < a \leq p(t) \leq b$. Then critical points of J will give us solutions of the initial boundary-value problem.

D - Nonlinear wave equation

For this we refer to [3], [16] and [17]. Note that one can also formulate along those lines a pair of conjugate variational principles for the sine-gordon equation, by writing it

$$u_{tt} - u_{xx} - u = -u + \sin u$$

where the left-hand side is a self-adjoint linear operator and the right is $-f'(u)$, with f convex.

4. Conclusion

In all the examples we have listed (except sine-gordon), the duality approach leads to existence results, because one actually shows that the dual functional has critical points.

We will not describe here this second step of the analysis. Let us simply mention that there are essentially two methods which can be used:

(a) the dual functional is coercive, weakly lower semi-continuous, and so has a global minimum (see [8]),

(b) the dual functional is unbounded, but yields to the mountain-pass theorem of Ambrosetti and Rabinowitz (see [11]).

Finally note that if one is interested in autonomous problems (i.e. the Hamiltonian or the potential does not depend explicitly on t), there usually is a trivial (constant) solution, which the analysis must somehow eliminate. This amounts to showing that the critical point we find for the dual functional is not the origin; we again refer to [8] or [11] to show how it is done.

BIBLIOGRAPHY

1. AUBIN-EKELAND, "Second-order evolution equations with convex Hamiltonian", Canadian Math. Bull. 23 (1980) p. 81-94.

2. AUCHMUTY, "Duality for non-convex variational principles", preprint, Indiana University, Bloomington, 1981.

3. BREZIS-CORON-NIRENBERG, "Free vibrations for a nonlinear wave equation and a theorem of P. Rabinowitz", Comm. Pure App. Math. 33 (1980), p.667-684.

4. CLARKE, "Solution périodique des équations hamiltoniennes", CRAS Paris, 287, 1978, p.951-2.

5. CLARKE, "Periodic solutions to Hamiltonian inclusions", J. Diff. Eq. 40, 1981, p.1-6.

6. CLARKE, "On Hamiltonian flows and symplectic transformations", SIAM J. Control and Optimization, in press.

7. CLARKE-EKELAND, "Solutions periodiques,de periode donnée, des équations hamiltoniennes", CRAS Paris 287, 1978, p.1013-1015.

8. CLARKE-EKELAND, "Hamiltonian trajectories having prescribed minimal period", Comm. Pure App. Math., 33 (1980), p.103-116.

9. CLARKE-EKELAND, "Nonlinear oscillations and boundary-value problems for Hamiltonian systems", Archive Rat. Mech. An., in press.

10. EKELAND, "Duality in non-convex optimization and calculus of variations", SIAM J. Opt. Con. 15, 1977, p.905-934.

11. EKELAND, "Periodic solutions to Hamiltonian equations and a theorem of P. Rabinowitz", J. Diff. Eq. 34, 1979, p.523-534.

12. EKELAND, "Forced oscillations for nonlinear Hamiltonian systems II", Advances in Mathematics, volume in honor of Laurent Schwartz, Nachbin ed., 1981, Academic Press.

13. EKELAND, "Oscillations forcées de systémes hamiltoniens non linéaires III", Bulletin de la SMF, in press.

14. EKELAND, "Dualité et stabilité des systémes hamiltoniens", preprint 1981.

15. EKELAND, "A perturbation theory near convex Hamiltonian systems", preprint, 1981.

16. EKELAND-LASRY, "Principes variationnels en dualité", CRAS Paris, 291 (1980), p.493-497.

17. EKELAND-LASRY, "Duality in nonconvex variational problems", preprint, CEREMADE, 1980.

18. EKELAND-LASRY, "On the number of closed trajectories for a Hamiltonian system on a convex energy surface", Annals of Math., 112 (1980), p.283-319.

19. EKELAND-TEMAM, "Convex analysis and variational problems", North-Holland-Elsevier, 1976.

20. TOLAND, "Stability of heavy rotating chains", J. Diff. Eq. 32 (1979), p.15-31.

21. TOLAND, "A duality principle for non-convex optimization and the calculus of variations", Arch. Rat. Mech. An. 71 (1979), p. 41-61.

22. TOLAND, "Duality in nonconvex optimization", J. Math. An. Appl. 66 (1978), p.41-61.

23. TRUC, Thése 3eme cycle, en cours, Université Paris-Dauphine.

Γ - CONVERGENCE AND CALCULUS OF VARIATIONS

ENNIO DE GIORGI GIANNI DAL MASO

1. DEFINITION AND ELEMENTARY EXAMPLES OF Γ - LIMITS.

In order to introduce in a natural way the definitions of the Γ - limits of a sequence of functions, we prefer to begin with some remarks about the definitions of *limit inferior* and *limit superior*.

Let f be a function, defined on a subset Y of a topological space X, and with values in the extended real line $\overline{\mathbb{R}} = \mathbb{R} \cup \{-\infty, +\infty\}$.
For every $x \in X$ we shall denote by $\mathcal{J}(x)$ the family of all neighbourhoods of x in X. for every $x \in \overline{Y}$ the limit inferior $\tilde{f}(x)$ of f is defined by

$$\tilde{f}(x) = \liminf_{y \to x} f(y) = \sup_{U \in \mathcal{J}(x)} \inf_{y \in U \cap Y} f(y).$$

We observe that the operator which transforms f into $\tilde{f}$ is in fact the product of two elementary operators: the first one maps the point-function f into the set-function g defined by

$$g(U) = \inf_{y \in U \cap Y} f(y)$$

the second one transforms the set-function g into the point-function $\tilde{f}$ defined by

$$\tilde{f}(x) = \sup_{U \in \mathcal{J}(x)} g(U).$$

An analogous decomposition is possible also for the limit superior, defined for every $x \in \overline{Y}$ by

$$\limsup_{y \to x} f(y) = \inf_{U \in \mathcal{J}(x)} \sup_{y \in U \cap Y} f(y).$$

In the particular case of a sequence $\{a_h\}$ of elements of $\overline{\mathbb{R}}$, we have

$$\liminf_{h \to \infty} a_h = \sup_{k \in \mathbb{N}} \inf_{h \geq k} a_h$$

$$\limsup_{h \to \infty} a_h = \inf_{k \in \mathbb{N}} \sup_{h \geq k} a_h$$

When f is a function of several variables, we can apply, for instance, first the limit superior with respect to one variable and then the limit inferior with respect to another. But, more generally, we can consider also some mixtures of half-limits with respect to one variable and half-limits with respect to the others. For example, if f is an extended-real valued function defined on the product $X \times Y$ of two topological spaces X and Y, we may consider, for every $x\in X$, $y\in Y$, the expression

$$\sup_{V\in\mathcal{J}(y)} \; \inf_{U\in\mathcal{J}(x)} \; \sup_{\eta\in V} \; \inf_{\xi\in U} \; f(\xi,\eta).$$

Expressions of this kind are called *hibrid limits* (see [30]). Some particular hibrid limits, named Γ - limits, are particularly useful in the study of sequences of problems in the calculus of variations.

We now recall the definition and the first elementary properties of the Γ - limits. Let $\{f_h\}$ be a sequence of functions defined on a topological space X and with values in $\overline{\mathbb{R}}$. For every $x\in X$ we define (see [28]):

$$\Gamma(X^-)\liminf_{\substack{h\to\infty\\ y\to x}} f_h(y) = \sup_{U\in\mathcal{J}(x)} \; \sup_{k\in\mathbb{N}} \; \inf_{h\geq k} \; \inf_{y\in U} f_h(y) = \sup_{U\in\mathcal{J}(x)} \; \liminf_{h\to\infty} \; \inf_{y\in U} f_h(y),$$

$$\Gamma(X^-)\limsup_{\substack{h\to\infty\\ y\to x}} f_h(y) = \sup_{U\in\mathcal{J}(x)} \; \inf_{k\in\mathbb{N}} \; \sup_{h\geq k} \; \inf_{y\in U} f_h(y) = \sup_{U\in\mathcal{J}(x)} \; \limsup_{h\to\infty} \; \inf_{y\in U} f_h(y)$$

$$\Gamma(X^+)\liminf_{\substack{h\to\infty\\ y\to x}} f_h(y) = \inf_{U\in\mathcal{J}(x)} \; \sup_{k\in\mathbb{N}} \; \inf_{h\geq k} \; \sup_{y\in U} f_h(y) = \inf_{U\in\mathcal{J}(x)} \; \liminf_{h\to\infty} \; \sup_{y\in U} f_h(y)$$

$$\Gamma(X^+)\limsup_{\substack{h\to\infty\\ y\to x}} f_h(y) = \inf_{U\in\mathcal{J}(x)} \; \inf_{k\in\mathbb{N}} \; \sup_{h\geq k} \; \sup_{y\in U} f_h(y) = \inf_{U\in\mathcal{J}(x)} \; \limsup_{h\to\infty} \; \sup_{y\in U} f_h(y)$$

REMARK 1.1. If the functions $f_h(x)$ are independent of x, that is $f_h(x) = a_h$ for every $x\in X$, then for every $x\in X$

$$\Gamma(X^-)\liminf_{\substack{h\to\infty\\ y\to x}} f_h(y) = \Gamma(X^+)\liminf_{\substack{h\to\infty\\ y\to x}} f_h(y) = \liminf_{h\to\infty} a_h,$$

$$\Gamma(X^-)\limsup_{\substack{h\to\infty\\ y\to x}} f_h(y) = \Gamma(X^+)\limsup_{\substack{h\to\infty\\ y\to x}} f_h(y) = \limsup_{h\to\infty} a_h.$$

If the functions $f_h(x)$ are independent of h, that is $f_h(x) = f(x)$ for every $h\in\mathbb{N}$, then for every $x \in X$

$$\Gamma(X^-)\liminf_{\substack{h\to\infty\\ y\to x}} f_h(y) = \Gamma(X^-)\limsup_{\substack{h\to\infty\\ y\to x}} f_h(y) = \liminf_{y\to x} f(y),$$

$$\Gamma(X^+)\liminf_{\substack{h\to\infty\\ y\to x}} f_h(y) = \Gamma(X^+)\limsup_{\substack{h\to\infty\\ y\to x}} f_h(y) = \limsup_{y\to x} f(y).$$

We say thet the sequence $\{f_h\}$ $\Gamma(X^-)$- converges to the limit $L\in\overline{\mathbb{R}}$ at the point $x\in X$, and write

$$\Gamma(X^-)\lim_{\substack{h\to\infty\\ y\to x}} f_h(y) = L,$$

if and only if

$$\Gamma(X^-)\liminf_{\substack{h\to\infty\\ y\to x}} f_h(y) = \Gamma(X^-)\limsup_{\substack{h\to\infty\\ y\to x}} f_h(y) = L.$$

Finally we say that $\{f_h\}$ $\Gamma(X^-)$ - converges to the function $f_\infty : X\to\overline{\mathbb{R}}$ if and only if

$$\Gamma(X^-)\lim_{\substack{h\to\infty\\ y\to x}} f_h(y) = f_\infty(x)$$

for every $x\in X$. In this case we use also the notations

$$\Gamma(X^-)\lim_{h\to\infty} f_h(x) = f_\infty(x)$$

or

$$\Gamma(X^-)\lim_{h\to\infty} f_h = f_\infty .$$

Similar conventions are adopted also with respect to the $\Gamma(X^+)$ - limits.

REMARK 1.2. For every sequence of functions $\{f_h\}$ and for every $x\in X$ we have

$$\Gamma(X^+)\liminf_{\substack{h\to\infty\\ y\to x}} f_h(y) = -\Gamma(X^-)\liminf_{\substack{h\to\infty\\ y\to x}} (-f_h(y)).$$

$$\Gamma(X^+)\limsup_{\substack{h\to\infty\\ y\to x}} f_h(y) = -\Gamma(X^-)\liminf_{\substack{h\to\infty\\ y\to x}} (-f_h(y)).$$

Therefore every property of the $\Gamma(X^+)$ - limits can be deduced from a corresponding property of the $\Gamma(X^-)$ - limits. This allows us to consider only the $\Gamma(X^-)$ - limits.

When the topological space X satisfies the first axiom of countability, in particular when X is metrizable, the $\Gamma(X^-)$ - limits can be chacacterized by means of sequences in X, as the following theorem shows.

THEOREM 1.3. Let x be an element of X with a countable neighbourhood base. Then

$$(1) \qquad L = \Gamma(X^-) \lim_{\substack{h \to \infty \\ y \to x}} f_h(y)$$

if and only if the following conditions are satisfied:

(a) for every sequence $\{x_h\}$ converging to x in X

$$L \leq \liminf_{h \to \infty} f_h(x_h),$$

(b) there exists a sequence $\{x_h\}$ converging to x in X such that

$$L = \lim_{h \to \infty} f_h(x_h).$$

Note that the implication (1) ⇒ (a) holds in every case, even if x has no countable neighbourhood base. In the case of a general topological space, condition (b) of the previous theorem is not always satisfied.

However conditions like (a) and (b) can be used to give a different definition of convergence, the sequential Γ - convergence, which is systematically studied in [42] .

We shall now compare the $\Gamma(X^-)$ - limit of a sequence of functions $\{f_h\}$ with the pointwise limit. First we shall give two elementary conditions, which ensure that the $\Gamma(X^-)$ - limit is equal to the pointwise limit. Then we shall give some examples where this is not true.

THEOREM 1.4. If $\{f_h\}$ is a sequence of lower semicontinuous functions, which converges uniformly to a function f_∞ , then

$$f_\infty = \Gamma(X^-) \lim_{h \to \infty} f_h.$$

THEOREM 1.5. If $\{f_h\}$ is an increasing sequence of lower semicontinuous functions, then for every $x \in X$

$$\lim_{h \to \infty} f_h(x) = \sup_{h \in \mathbb{N}} f_h(x) = \Gamma(X^-) \lim_{h \to \infty} f_h(x).$$

We remark that the $\Gamma(X^-)$ - limit is, in general, different from the pointwise limit. For instance, if $X = \mathbb{R}$ and

$$f_h(x) = \operatorname{arctg} [(hx-1)^2] ,$$

then we have

$$\Gamma(X^-) \lim_{\substack{h\to\infty \\ y\to x}} f_h(y) = \begin{cases} 0 & \text{if } x = 0, \\ \frac{\pi}{2} & \text{if } x \neq 0, \end{cases}$$

whereas

$$\lim_{h\to\infty} f_h(x) = \begin{cases} \frac{\pi}{4} & \text{if } x = 0, \\ \frac{\pi}{2} & \text{if } x \neq 0. \end{cases}$$

In some cases the $\Gamma(X^-)$ - limit exists, but the poinwise limit does not exist. For instance, if $X = \mathbb{R}$ and $f_h(x) = \sin(hx)$, then

$$\Gamma(X^-) \lim_{h\to\infty} f_h = -1,$$

but $\lim_{h\to\infty} f_h(x)$ does not exist, unless $\frac{x}{\pi}$ is an integer.

In some other cases the pointwise limit exists, whereas the $\Gamma(X^-)$ - limit does not exist. For instance, if $X = \mathbb{R}$ and

$$f_h(x) = \begin{cases} (-1)^h [\sin(h!\, 2\pi x)]^2 & \text{if } x \text{ is rational,} \\ 0 & \text{if } x \text{ is irrational,} \end{cases}$$

then for every $x \in X$

$$\Gamma(X^-) \liminf_{\substack{h\to\infty \\ y\to x}} f_h(y) = -1,$$

$$\Gamma(X^-) \limsup_{\substack{h\to\infty \\ y\to x}} f_h(y) = 0,$$

$$\lim_{h\to\infty} f_h(x) = 0.$$

2 SOME ABSTRACT PROPERTIES.

We now state some abstract properties of Γ - limits, that are particularly useful for the calculus of variations. For a complete exposition of all abstract properties of Γ - limits see [28] , for the proofs see [29] .

THEOREM 2.1. If X has a countable base, then for every sequence $\{f_h\}$ of functions from X into $\overline{\mathbb{R}}$ there exists a subsequence $\{f_{h_k}\}$ and a function $f_\infty : X \longrightarrow \overline{\mathbb{R}}$ such that

$$f_\infty = \Gamma(X^-) \lim_{k \to \infty} f_{h_k} .$$

THEOREM 2.2. If $\{f_h\}$ $\Gamma(X^-)$ - converges to f_∞ , and if $\{x_h\}$ is a sequence in X converging to x_∞ with

$$\liminf_{h \to \infty} f_h(x_h) = \liminf_{h \to \infty} \inf_{x \in X} f_h(x),$$

then

$$f_\infty(x_\infty) = \min_{x \in X} f_\infty(x) = \lim_{h \to \infty} \inf_{x \in X} f_h(x)$$

The preceding theorem applies for instance when each x_h is a minimum point of f_h and $\{x_h\}$ converges to x_∞ .

THEOREM 2.3. If $\{g_h\}$ is a sequence of continuous functions from X into $\mathbb{R}$, which converges uniformly to a function g_∞ , then for every $x \in X$

$$\Gamma(X^-) \liminf_{\substack{h \to \infty \\ y \to x}} (f_h + g_h)(y) = \Gamma(X^-) \liminf_{\substack{h \to \infty \\ y \to x}} f_h(y) + g_\infty(x)$$

$$\Gamma(X^-) \limsup_{\substack{h \to \infty \\ y \to x}} (f_h + g_h)(y) = \Gamma(X^-) \limsup_{\substack{h \to \infty \\ y \to x}} f_h(y) + g_\infty(x)$$

THEOREM 2.4. Let f'_∞ and f''_∞ be the functions defined by

$$f'_\infty(x) = \Gamma(X^-) \liminf_{\substack{h \to \infty \\ y \to x}} f_h(y),$$

$$f''_\infty(x) = \Gamma(X^-) \limsup_{\substack{h \to \infty \\ y \to x}} f_h(y) ;$$

then f'_∞ and f''_∞ are lower semicontinuous on X; moreover for every $x \in X$ we have

$$f'_\infty(x) = \Gamma(X^-) \liminf_{\substack{h\to\infty \\ y\to x}} (sc^- f_h)(y),$$

$$f''_\infty(x) = \Gamma(X^-) \limsup_{\substack{h\to\infty \\ y\to x}} (sc^- f_h)(y),$$

where $sc^- g$ denotes the greatest lower semicontinuous function majorized by g.

If $f_h = f$ for every $h \in \mathbb{N}$, then $f'_\infty = f''_\infty = sc^- f$.

DEFINITION 2.5. We say that a sequence $\{f_h\}$ is equicoercive if, for every $t \in \mathbb{R}$, there exists a sequentially compact subset K_t of X such that

$$\{x \in X : f_h(x) \leqslant t\} \subset K_t$$

for every $h \in \mathbb{N}$.

THEOREM 2.6. Let $\{f_h\}$ be an equicoercive sequence of functions from X into $\overline{\mathbb{R}}$, which $\Gamma(X^-)$ - converges to f_∞, and let $\{g_h\}$ be a sequence of non - negative continuous functions from X into $\mathbb{R}$, which converges uniformly to a function g_∞.

Then

$$\min_{x\in X} [f_\infty(x) + g_\infty(x)] = \lim_{h\to\infty} \inf_{x \in X} [f_h(x) + g_h(x)] .$$

If, in addition, $f_\infty + g_\infty$ has a unique minimum point x_∞, and if $\{x_h\}$ is a sequence in X such that each x_h is a minimum point of $f_h + g_h$, then the sequence $\{x_h\}$ converges to x_∞ in X.

REMARK 2.7. Theorem 2.6. plays a fundamental role in the applications of Γ- convergence to the calculus of variations. In many cases it allows to infer, from the $\Gamma(X^-)$ - convergence of the sequence $\{f_h\}$, the convergence of the minimum values and of the minimum points of all sequences of the form $f_h + g$, with g continuous and bounded from below.

REMARK 2.8. In many problems of the calculus of variations it is interesting to prove also the converse of theorem 2.6., that is to infer the equality

$$f_\infty = \Gamma(X^-) \lim_{h\to\infty} f_h$$

from the fact that

$$\inf_{x\in X} [f_\infty(x) + g(x)] = \lim_{h\to\infty} \inf_{x\in X} [f_h(x) + g(x)]$$

for every g in a suitable class of test functions.
The following theorem is a result in this direction.

THEOREM 2.9. Let (X, d) be a metric space and let $\{f_h\}$ be a sequence of functions from X into $[0, +\infty]$. For every $\alpha>0$ and for every $x\in X$ we have

$$\Gamma(X^-) \liminf_{\substack{h\to\infty \\ y\to x}} f_h(y) = \sup_{\lambda>0} \liminf_{h\to\infty} \inf_{y\in X} [f_h(y) + \lambda d(x,y)^\alpha] ,$$

$$\Gamma(X^-) \limsup_{\substack{h\to\infty \\ y\to x}} f_h(y) = \sup_{\lambda>0} \limsup_{h\to\infty} \inf_{y\in X} [f_h(y) + \lambda d(x,y)^\alpha] .$$

In particular, if there exist $\alpha>0$ and a lower semicontinuous function $f_\infty : X\to[0, +\infty]$ such that

$$\inf_{y\in X} [f_\infty(y) + \lambda d(x,y)^\alpha] = \lim_{h\to\infty} \inf_{y\in X} [f_h(y) + \lambda d(x,y)^\alpha]$$

for every $x\in X$ and for every $\lambda>0$, then

$$f_\infty = \Gamma(X^-) \lim_{h\to\infty} f_h$$

3. INDIRECT METHODS IN THE STUDY OF LIMITS OF VARIATIONAL PROBLEMS.

We now describe the situation, that most frequently occurs in those applications of the Γ - convergence to the calculus of variations, which have been studied till now.

Let Ω be an open subset of $\mathbb{R}^n$, let X be a topological vector space of real functions defined on Ω, and let Y be a subset of X contained in the Sobolev space $W^{1,1}_{loc}(\Omega)$. Let $F_h : X \to \overline{\mathbb{R}}$ be a sequence of integral functionals of the form:

$$F_h(u) = \begin{cases} \int_\Omega f_h(x,u(x), Du(x))\, dx & \text{if } u\in Y \\ +\infty & \text{if } u\in X - Y, \end{cases}$$

where Du denotes the distributional gradient of u and f_h is a sequence of non-negative

Borel functions defined on $\Omega \times \mathbb{R} \times \mathbb{R}^n$ and with values in $\overline{\mathbb{R}}$. We shall suppose that the sequence of functionals $\{F_h\}$ is equicoercive and $\Gamma(X^-)$ - converges to a functional $F_\infty : X \to \overline{\mathbb{R}}$. For every $h \in \mathbb{N}$ let $\overline{F}_h = sc^- F_h$ be the lower semicontinuous envelope of F_h. The theorems stated in the preceding section enable to infer that

$$(1)\quad \lim_{h \to \infty} \inf_{u \in X} [F_h(u) + G(u)] = \lim_{h \to \infty} \min_{u \in X} [\overline{F}_h(u) + G(u)] = \min_{u \in X} [F_\infty(u) + G(u)]$$

for every $G : X \to \mathbb{R}$ continuous and non-negative. In many cases G also is an integral functional of the form

$$G(u) = \int_\Omega g(x,u(x))\, dx,$$

with g sufficiently regular, in order to have the continuity of the functional G with respect to the same topology, in which the Γ - limit operation is carried out.

From (1) it follows that, if we are able to identify the functional F_∞, we can also determine the limits of many variational problems related to the functionals F_h.

In order to describe the functional F_∞, it is important to answer the following questions.

1. Does there exist a Borel function $f_\infty : \Omega \times \mathbb{R} \times \mathbb{R}^n \to [0, +\infty]$ such that

   $$F_\infty(u) = \int_\Omega f_\infty(x, u(x), Du(x))\, dx$$

   for every $u \in Y$?

2. Is it possible to compute f_∞ explicitly, or, at least, to determine some of its properties?

When $X = L^p(\Omega)$, in some cases we can answer these questions by using the following *indirect method:* we determine the minimum values of the functionals

$$F_h(u) + \lambda \int_\Omega |u(x) - g(x)|^p\, dx$$

by solving their Euler equations, then we take the limit as h tends to $+\infty$, finally we determine F_∞ by means of theorem 2.9.

The crucial point of the indirect method is usually the determination of the limits of the solutions of some sequences of differential equations.

We now give an example, in which the indirect method can be applied rather easily.

EXAMPLE 3.1. Let $\Omega =]0,1[$, let $\{a_h\}$ be a sequence in $L^\infty(\Omega)$ and let $a_\infty \in L^\infty(\Omega)$; assume that there exist two constants c_1 and c_2, with $0 < c_1 \leq c_2 < +\infty$, such that $c_1 \leq a_h(x) \leq c_2$ a.e. on $]0,1[$ for every $h \in \mathbb{N} \cup \{\infty\}$

Let $X = L^2(\Omega)$ and let $Y = \{u \in W^{1,2}(\Omega) : u(0) = 0\}$; for every $u \in X$, $h \in \mathbb{N} \cup \{\infty\}$ we def

$$F_h(u) = \begin{cases} \int_\Omega a_h(x) \, |u'(x)|^2 \, dx & \text{if } u \in Y, \\ +\infty & \text{if } u \notin Y. \end{cases}$$

If $\{\frac{1}{a_h}\}$ converges to $\frac{1}{a_\infty}$ weakly in $L^1(\Omega)$, then

$$F_\infty = \Gamma(X^-) \lim_{h \to \infty} F_h .$$

PROOF: We fix $g \in L^2(\Omega)$ and $\lambda > 0$. For every $h \in \mathbb{N} \cup \{\infty\}$ we denote by u_h the unique minimum point of the functional

$$F_h(u) + \lambda \int_\Omega (u - g)^2 \, dx ,$$

which is the unique solution of the Euler equation

$$(2) \quad \begin{aligned} &(a_h u')' = \lambda (u - g) \\ &u(0) = 0 \\ &u'(1) = 0 . \end{aligned}$$

Since

$$c_1 \int_\Omega u'^2_h \, dx \leq F_h(u_h) + \lambda \int_\Omega (u_h - g)^2 \, dx \leq \lambda \int_\Omega g^2 \, dx,$$

the sequence $\{u_h\}$ is relatively compact in $L^2(\Omega)$. Let $\{u_{h_k}\}$ be a subsequence of $\{u_h\}$ which converges in $L^2(\Omega)$ to a function v_∞ . Since the Euler equation (2) is equivale to the integral equation

$$(3) \quad u(t) = \lambda \int_0^t \{\frac{1}{a_h(s)} \int_s^1 [g(\sigma) - u(\sigma)] \, d\sigma\} \, ds,$$

from the strong convergence of $\{u_{h_k}\}$ and the weak convergence of $\{\frac{1}{a_{h_k}}\}$ it follows that v_∞ is a solution of the integral equation (3) with $h = \infty$, hence $v_\infty = u_\infty$. This implies that $\{u_h\}$ converges to u_∞ in $L^2(\Omega)$. Equation (2) and an integration by parts give for every $h \in \mathbb{N} \cup \{\infty\}$

$$\min_{u \in X} [F_h(u) + \lambda \int_\Omega (u - g)^2 \, dx] = F_h(u_h) + \lambda \int_\Omega (u_h - g)^2 \, dx = \lambda \int_\Omega g(g - u_h) \, dx,$$

hence

$$\min_{u \in X} [F_\infty(u) + \lambda \int_\Omega (u - g)^2 \, dx] = \lim_{h \to \infty} \min_{u \in X} [F_h(u) + \lambda \int_\Omega (u - g)^2 \, dx] .$$

Since this is true for every $g \in L^2(\Omega)$ and for every $\lambda > 0$, by theorem 2.9. we have

$$(4) \qquad F_\infty = \Gamma(X^-) \lim_{h \to \infty} F_h .$$

REMARK 3.2. The knowledge of (4) and the abstract theory of section 2 enable us to prove many results that can not be obtained directly from the convergence of the corresponding differential equations. For instance, from (4) and from theorem 2.6. it follows that

$$(5) \qquad \min_{u(0)=0} \int_0^1 \{a_\infty |u'|^2 + g(x, u)\} dx = \lim_{h\to\infty} \min_{u(0)=0} \int_0^1 \{a_h |u'|^2 + g(x,u)\} dx,$$

whenever the function $g(x,t)$ is measurable in x, continuous in t, and satisfies

$$0 \leq g(x,t) \leq c_3 |t|^2 + c_4$$

for some constants $c_3, c_4 \in \mathbb{R}$. We remark that, even if $g(x,t)$ is differentiable in t, it is often difficult to obtain (5) working only with differential equations, without any reference to Γ- convergence.

The indirect method can be employed also in the study of the Γ-limits of sequences of equi-uniformly elliptic quadratic functionals of the form

$$F_h(u) = \int_\Omega \{ \sum_{ij=1}^n a_{ij}^{(h)}(x) \, D_i u(x) D_j u(x) \} dx.$$

In this case we can state a connection between the Γ-limit of the sequence $\{F_h\}$ and the G-limit of the sequence of the corresponding second order elliptic differential operators (see [51] , [33] , [7])

We observe that the arguments of the paper [33] (written before [28] , thus before the first abstract treatment of the notion of $\Gamma(X^-)$ -limit) may be considered as an anticipation of the indirect method described in this section.

Further applications of the indirect method can be obtained from the large number of recents results in the theory of the asymptotic limits of solutions of differential equations. Recent general expositions of the results obtained in this field have been written by A. Bensoussan, J.L. Lions, G. Papanicolaou [7] , J.L. Lions [36] , E. Sanchez-Palencia [46] , L. Tartar [52] , V.V. Zhikov, S.M. Kozlov, O.A. Oleinik, Kha T'en Ngoan [54]

These surveys contain many important examples of problems that are interesting from the point of view both of pure analysis and of applied mathematics and numerical calculus; we point out, among them, the results about homogenization, which corresponds to the study of composite materials and porous media.

4. DIRECT METHODS IN Γ- CONVERGENCE.

In many problems of the calculus of variations it is impossible to use the indirect methods, which consists essentially in solving a sequence of Euler equations and in studying the limits of the solutions. In many cases, however, we are still able to prov some properties of the Γ-limit, and sometimes to compute it explicitly, by using some different techniques, that have developed in recent years, and which we call direct met in Γ-convergence.

These methods provide the most significant applications of the Γ-convergence to the calculus of variations. We remark that the direct methods employ in an essential manner the results of the abstract theory of Γ-convergence, and that, at least at the beginni this theory has developed in order to study the limits of those variational problems, that can not be attacked by indirect methods.

The paper [24] can be considered as an anticipation of the direct methods in Γ-convergence. Afterwards these methods have been applied by many authors in a systematic wa

The main idea of the direct method is to study the Γ-limit not for a single open set Ω, but for all open subsets of $\mathbb{R}^n$, or for a sufficiently large class of open sets

Given a sequence of integrands $\{f_h\}$, and the corresponding functionals

$$F_h(u, A) = \int_A f_h(x,u,Du)\,dx,$$

we study the functional

$$F_\infty(u, A) = \Gamma(X_A^-) \lim_{\substack{h\to\infty \\ v\to u}} F_h(v, A),$$

where A varies in a suitable class of open sets and, for every A, the function u varies in a suitable class of functions X_A. First we examine the dependence of the functional F_∞ on the variable A, in order to prove that the set function $A \to F_\infty(u, A)$ is the tra of a measure. Once this results has been established, it remains to prove that this measure can be written in the form

$$F_\infty(u, A) = \int_A f_\infty(x,u,Du)\,dx.$$

In this second part an important role is played by the properties of the functional F_∞ with respect to the variable u, as, for instance, continuity or convexity.

The results of the following example have been proved by these methods.

EXAMPLE 4.1. (See [47]). Let $\{f_h\}$ be a sequence of Borel functions, defined on $\mathbb{R}^n \times \mathbb{R}$ and with values in $\mathbb{R}$, and let $1 \leq p < +\infty$. Assume that there exists a constant $c > 0$

such that

$$(a)\quad |z|^p \leq f_h(x,s,z) \leq c(1 + |s|^p + |z|^p),$$

$$(b)\quad \left| f_h(x,s_1,z_1)^{\frac{1}{p}} - f_h(x, s_2, z_2)^{\frac{1}{p}} \right| \leq c\,(|s_1 - s_2| + |z_1 - z_2|).$$

Then there exist a subsequence $\{f_{h_k}\}$ of $\{f_h\}$ and a Borel function f_∞ satisfying (a) and (b), such that for every bounded open subset A of $\mathbb{R}^n$ and for every $u \in L^p(A)$

$$\int_A f_\infty(x,u,Du)\,dx = \Gamma(L^p(A)^-) \lim_{\substack{k\to\infty \\ v\to u}} \int_A f_{h_k}(x,v,Dv)\,dx$$

(the integrals are assumed to be $+\infty$ when $u \notin W^{1,p}(A)$).

In the case p = 2, if the integrands f_h have the form

$$f_h(x,s,z) = \sum_{ij=1}^{n} a_{ij}^{(h)}(x)\, z_i z_j\,,$$

then also f_∞ has the form

$$f_\infty(x,s,z) = \sum_{ij=1}^{n} a_{ij}^{(\infty)}(x)\, z_i z_j.$$

For other results of this kind, obtained by weakening the hypotheses (a) or (b), we refer to [37], [17], [16], [10].

Besides these results there are problems in which there is a drastic change in the form of the integrands, as the following examples show.

EXAMPLE 4.2. (See [9]). Let $f : \mathbb{R} \to \mathbb{R}$ be the function defined by $f(r) = |r| \min \{\int_0^{\frac{1}{|r|}} [|v'(s)|^2 + \sin(2\pi\, v(s))]\, ds : v(0) = 0,\ v(\frac{1}{|r|}) = 1\}$ for every $r \neq 0$, and by $f(0) = -1$ for s = 0. Then for every bounded open subset A of $\mathbb{R}^n$ and for every $u \in L^2(A)$

$$\int_A f(|Du|)\,dx = \Gamma(L^2(A)^-) \lim_{\substack{h\to\infty \\ v\to u}} \int_A [|Du|^2 + \sin(2\pi\, hu)]\,dx$$

(the integrals are assumed to be $+\infty$ if $u \notin W^{1,2}(A)$).

The function f is strictly convex on $\mathbb{R}$, analytic in $\mathbb{R} - \{0\}$, and satisfies the following conditions:

$$(a)\quad |r|^2 - 1 \leq f(r) \leq |r|^2 \quad \text{for every } r \in \mathbb{R}$$

$$(b)\quad \lim_{r\to\infty} [f(r) - |r|^2] = 0$$

$$(c)\quad \lim_{r\to 0^+} \frac{f(r) - f(0)}{r} = \frac{4\sqrt{2}}{\pi}$$

Therefore in this example the $\Gamma(L^2(A)^-)^-$-limit of a sequence of integral functior quadratic in Du, is *not quadratic* in Du.

The same phenomenon occurs in the following example.

EXAMPLE 4.3. (See [14]). Let $f : \mathbb{R}^n \times \mathbb{R}^n \to \mathbb{R}$ be a Borel function, bounded on the bounded sets, with the following properties:

(a) $f(x,z) \geq 0$ for every $x \in \mathbb{R}^n$, $z \in \mathbb{R}^n$,

(b) for every $x \in \mathbb{R}^n$ the function $z \to f(x,z)$ is convex on $\mathbb{R}^n$

(c) for every $z \in \mathbb{R}^n$ the function $x \to f(x,z)$ is periodic with period 1 in each variable $x_1, \dots, x_n$.

For every bounded open subset A of $\mathbb{R}^n$ and for every $u \in L^2(A)$ we set

$$F_h(u, A) = \begin{cases} \int_A f(hx, du)\, dx & \text{if } |Du| \leq 1 \text{ a.e. on } A \\ +\infty & \text{otherwise} \end{cases}$$

Let us denote by P the unit cube $[0,1]^n$; for every $z \in \mathbb{R}^n$ with $|z| \leq 1$ we denote by K(z) the set of all lipschitzian functions $u : \mathbb{R}^n \to \mathbb{R}$ such that

(d) $\int_P Du\,dx = z$

(e) $|Du| \leq 1$ a. e. $\mathbb{R}^n$

(f) Du is periodic of period 1 in each variable $x_1, \dots, x_n$.

For every $z \in \mathbb{R}^n$, with $|z| \leq 1$, we set

$$g(z) = \min_{u \in K(z)} \int_P f(x, Du)\, dx;$$

for every bounded open subset A of $\mathbb{R}^n$ and for every $u \in L^2(A)$ we set

$$G(u, A) = \begin{cases} \int_A g(Du)\, dx & \text{if } |Du| \leq 1 \text{ a. e. on } A \\ +\infty & \text{otherwise} \end{cases}$$

Then for every bounded open subset A of $\mathbb{R}^n$ and for every $u \in L^2(A)$

$$G(u, A) = \Gamma(L^2(A)^-) \lim_{\substack{h \to \infty \\ v \to u}} F_h(v, A).$$

If f is a quadratic form in z, i. e.

$$f(x,z) = \sum_{ij=1}^{n} a_{ij}(x)\, z_i z_j,$$

then the function g *is not*, in general, *a quadratic form* in z. For example, if $n = 1$, $0 < \alpha < \beta$, and

$$f(x,z) = \begin{cases} \alpha z^2 & \text{if } k \leq x < k + \frac{1}{2}, \quad k \in Z \\ \beta z^2 & \text{if } k - \frac{1}{2} \leq x < k, \quad k \in Z \end{cases}$$

then (see [13])

$$g(z) = \begin{cases} \left[\frac{1}{2}\left(\frac{1}{\alpha} + \frac{1}{\beta}\right)\right]^{-1} z^2 & \text{if } |z| \leq \alpha\left[\frac{1}{2}\left(\frac{1}{\alpha} + \frac{1}{\beta}\right)\right] \\ \frac{\alpha}{2} + \frac{\beta}{2}(2z-1)^2 & \text{if } \alpha\left[\frac{1}{2}\left(\frac{1}{\alpha} + \frac{1}{\beta}\right)\right] \leq |z| \leq 1 . \end{cases}$$

Finally there are some examples in which the Γ-limit F_∞ (u, A) is a measure with respect to A, but it can not be written as an integral with respect to the Lebesgue measure.

EXAMPLE 4.4. (See [41]). For every bounded open subset A of $\mathbb{R}^n$ and for every $u \in L^1(A)$ we set

$$F_h(u, A) = \begin{cases} \int_A \left[\frac{|Du|^2}{h} + h(\cos u + 1)\right] dx & \text{if } Du \in L^2(A), \\ +\infty & \text{otherwise,} \end{cases}$$

$$\int_A |Du| = \sup \{ \int_A u \operatorname{div} g dx : g \in [C_o^1(A)]^n, |g| \leq 1 \}$$

$$F_\infty(u,A) = \begin{cases} \frac{4\sqrt{2}}{\pi} \int_A |Du| & \text{if } \cos u = -1 \text{ a.e. on } A, \\ +\infty & \text{otherwise.} \end{cases}$$

Then for every bounded open subset A of $\mathbb{R}^n$, with a Lipschitzian boundary, and for every $u \in L^1(A)$ we have

$$F_\infty(u, A) = \Gamma(L^1(A)^-) \lim_{\substack{h \to \infty \\ v \to u}} F_h(v, A).$$

In this example a connection has been stated between two different kinds of functionals: the integral functionals F_h, elliptic and with quadratic principal part, whose Euler equations are related to the equation $\Delta u + \sin u = 0$, and the "total variation" $\int_A |Du|$, which is related to the theory of minimal surfaces (see [26]).

In the following example, for every $u \in W^{1,2}(A)$ we shall denote by $\tilde{u}$ the quasi continuous representative of u, which is defined up to sets of capacity 0. A property is said to hold capacity almost everywhere on A (abbreviated cap. a.e. on A) if the set of points of A where it fails to hold is a set of capacity 0.

EXAMPLE 4.5. (See [27], [23], [22]). Let $\{\varphi_h\}$ and $\{\Psi_h\}$ be two sequences of functions defined on $\mathbb{R}^n$ and with values in $\overline{\mathbb{R}}$. Suppose that there exists a sequence $\{w_h\}$, bounded in $W^{1,2}_{loc}(\mathbb{R}^n)$, such that $\varphi_h \leq \tilde{w}_h \leq \Psi_h$ cap. a.e. on $\mathbb{R}^n$. For every bounded open subset A of $\mathbb{R}^n$ and for every $u \in L^2(A)$ we set

$$F_h(u,A) = \begin{cases} \int_A |Du|^2 dx & \text{if } u \in W^{1,2}(A) \text{ and } \varphi_h \leq \tilde{u} \leq \Psi_h \text{ cap. a.e. on } A, \\ +\infty & \text{otherwise.} \end{cases}$$

Suppose that there exists a functional $F_\infty(u, A)$, increasing in A, with the following properties:

(a) for every bounded open set A and for every $u \in L^2(A)$

$$F_\infty(u, A) = \sup\{ F_\infty(u, A') : A' \text{ open}, A' \subset\subset A\},$$

(b) for every pair (A_1, A_2) of bounded open sets, with $A_1 \subset\subset A_2$, there exists an open set A, with $A_1 \subset\subset A \subset\subset A_2$, such that

$$F_\infty(u, A) = \Gamma(L^2(A)^-) \lim_{\substack{h \to \infty \\ v \to u}} F_h(v, A)$$

for every $u \in L^2(A)$.

Then the functional F_∞ can be written as an integral in the following form:

$$F_\infty(u, A) = \begin{cases} \int_A |Du|^2 dx + \int_A f(x, \tilde{u}(x))\, d\mu(x) + \gamma(A) & \text{if } u \in W^{1,2}(A), \\ +\infty & \text{otherwise,} \end{cases}$$

where μ, γ are positive Radon measures, $\mu \in W^{-1,2}(\mathbb{R}^n)$, and $f : \mathbb{R}^n \times \mathbb{R} \to [0,+\infty]$ is a Borel function such that for every $x \in \mathbb{R}^n$ the function $t \to f(x,t)$ is convex and lower semicontinuous on $\mathbb{R}$.

Moreover a compactness theorem ensures that for every sequence $\{F_h\}$ there exist a subsequence and a functional F_∞, which satisfy (a) and (b).

In some cases the additional term

$$(1) \quad \int_A f(x, \tilde{u}(x)) d\mu(x) + \gamma(A)$$

takes only the values 0 and $+\infty$; in this case the limit problem is still an obstacle problem; indeed there exist two functions $\varphi, \Psi : \mathbb{R}^n \to \overline{\mathbb{R}}$ such that for every bounded open set A and for every $u \in W^{1,2}(A)$

$$\int_A f(x, \tilde{u}(x))\, d\mu(x) + \gamma(A) = \begin{cases} 0 & \text{if } \varphi \leq \tilde{u} \leq \Psi \text{ cap. a.e. on } A \\ +\infty & \text{otherwise} \end{cases}$$

Nevertheless there are also some cases, in which the functional (1) takes every positive real value. For instance (see [15]), if n = 2 and

$$\Psi_h(x) = -\varphi_h(x) = \begin{cases} 0 & \text{if dist } (x, \frac{1}{h} Z^2) \leq \exp(-h^2), \\ +\infty & \text{otherwise,} \end{cases}$$

($\frac{1}{h} Z^2$ denotes the set of all points of the form $(\frac{m}{h}, \frac{n}{h})$, with $(m,n) \in Z^2$), then

$$F_\infty(u, A) = \begin{cases} \int_A |Du|^2 dx + 2\pi \int_A u^2\, dx & \text{if } u \in W^{1,2}(A), \\ +\infty & \text{otherwise.} \end{cases}$$

We refer to [15], [5], [19][43] for an explicit computation of the functional F_∞ for certain particular sequences of obstacles $\{\varphi_h\}$ and $\{\Psi_h\}$.

It is not possible to expose here the techniques used in the proofs of the results described in these examples. We point out, however, that all abstract theorems of the theory of Γ -convergence play an important role in these proofs.

We can say that the direct methods, whose results are illustrated by the examples of this section, can be considered for the present as the main source of motivation and the largest field of application for the abstract theory of Γ -convergence, outlined in [28] and [8].

5. SOME OPEN PROBLEMS.

There are many open problems in the theory of the limits of variational functionals. We point out, in particular, the study of limits of integrals depending on higher order derivatives, or on the first derivatives of vector valued functions. Moreover it would be interesting to study the limits of functionals of the following form (frequent in control theory):

$$F(u, v, \Omega) + G(u, v, \Omega),$$

where $F(u, v, \Omega)$ is an integral functional (like those considered in the preceding section) and $G(u, v, \Omega) = 0$ if a particular differential equation involving u and v is satisfied on , $G(u, v, \Omega) = +\infty$ otherwise.

We remark that, besides the study of limits of minimum points we can also consider the problem of the limits of stationary points, and, more generally, of the limits of the solutions of functional equations of the form

(1) $\operatorname{grad} f_h(u) = v,$

(2) $\frac{du}{dt} = - \operatorname{grad} f_h(u(t)),$

(3) $\frac{d^2u}{dt^2} = - \operatorname{grad} f_h(u(t)),$

where f_h is a sequence of functions defined on a Hilbert space and with values in $\overline{\mathbb{R}}$. A number of problems of elliptic, parabolic, hyperbolic type can be reduced to this form.

The first papers devoted to the study of (1) and (2), for non-convex functions, are [1], [31], [38], [39].

The most recent studies in this field lead to introduce a new generalization of convex functions: (p, q) - convex functions. We recall the definition (see [32]): let $p, q \in \mathbb{R}$ and let f be a lower semicontinuous function defined on a Hilbert space H and with values in $]-\infty, +\infty]$; we say that f is (p, q) -convex if for every pair (x, y) of elements of H there exists $z \in H$ such that

$$z - \frac{x + y}{2} \leq p\,|x - y|^2$$

$$f(z) \leq \frac{f(x) + f(y)}{2} - q\,|x - y|^2$$

The systematic study of (3) seems to be more complex and difficult, even in the convex case. In order to indicate the difficulties that arise in such problems, we refer to the following papers: [45] for a finite dimensional case; [2][3], [4], [20], [21], [48], [49], [50] for the problem of a string vibrating against an obstacle; [18], [11], [12] for the elastic bounce problem.

Finally there are the problems of stochastic Γ-convergence, that is the study of Γ-limits of integral functionals which depend on random integrands. In recent years many results have been proved in the theory of stochastic homogenization, for which we refer to [6], [34], [35], [40], [44], [53], [54]. The methods used by these authors can be compared with the indirect methods described in section 3.

In order to study the theory of stochastic homogenization, and other problems of this kind, by means of the direct methods of Γ-convergence, we propose here a possible definition of convergence for a sequence of probability measures defined on spaces of lower semicontinuous functions.

Let X be a topological space; we denote by S the set of all lower semicontinuous functions defined on X and with values in $\overline{\mathbb{R}}$.

A subset A of S is said to be sequentially $\Gamma(X^-)$ -open if, for every $f \in A$ and for every sequence $\{f_h\}$ in S, which $\Gamma(X^-)$ -converges to f, there exists $k \in \mathbb{N}$ such that $f_h \in A$ for every $h \geq k$.

A subset K of S is said to be sequentially $\Gamma(X^-)$ -compact if, for every sequence $\{f_h\}$ in K, there exists a subsequence $\{f_{h_k}\}$ which $\Gamma(X^-)$ -converges to an element f of K.

Let us denote by $\mathcal{A}$ the class of all sequentially $\Gamma(X^-)$ -open subsets of S and by $\mathcal{K}$ the class of all sequentially $\Gamma(X^-)$ -compact subsets of S.

Let $\{\mu_h\}$ be a sequence of increasing functions, defined on the set $P(S)$ of all subsets of S, and with values in $[0,+\infty]$ (for instance a sequence of probability measures on S). For every $E \subseteq S$ we define

$$\tilde{\Gamma}(X^-) \liminf_{h\to\infty} \mu_h(E) = \inf_{E \subset \in \mathcal{A}} \sup_{A \supset K \in \mathcal{K}} \liminf_{h\to\infty} \mu_h(K),$$

$$\tilde{\Gamma}(X^-) \limsup_{h\to\infty} \mu_h(E) = \inf_{E \subseteq A \in \mathcal{A}} \sup_{A \supset K \in \mathcal{K}} \limsup_{h\to\infty} \mu_h(K).$$

We say that $\{\mu_h\}$ $\tilde{\Gamma}(X^-)$ -converges to the increasing function $\mu_\infty : P(S) \to [0,+\infty]$ if for every $E \subseteq S$

$$\tilde{\Gamma}(X^-) \liminf_{h\to\infty} \mu_h(E) = \tilde{\Gamma}(X^-) \limsup_{h\to\infty} \mu_h(E) = \mu_\infty(E).$$

It would be interesting to find reasonable hypotheses under which the following property holds (note the analogy with theorem 2.6.).

PROPERTY 5.1. If $\{\mu_h\}$ $\tilde{\Gamma}(X^-)$ -converges to μ_∞, then for every non-negative continuous function $g : X \to [0, +\infty[$ we have

$$\int_S \min_{x\in X} [f(x) + g(x)]\, d\mu_\infty(f) = \lim_{h\to\infty} \int_S \min_{x\in X} [f(x) + g(x)]\, d\mu_h(f),$$

where the integral of a non-negative function $G : S \to [0,+\infty]$ with respect to an increasing function $\lambda : P(S) \longrightarrow [0,+\infty]$ is defined by

$$\int_S G(f)\, d\lambda(f) = \int_0^{+\infty} \lambda(\{f \in S : G(f) > t\})\, dt.$$

It would also be interesting to give sufficient conditions on $\{\mu_h\}$ in order that there exists an element f_∞ of S such that

$$\mu_\infty(E)=\delta_{f_\infty}(E)=\begin{cases}1 & \text{if } f_\infty \in E,\\ 0 & \text{if } f_\infty \notin E.\end{cases}$$

R E F E R E N C E S

[1] A. AMBROSETTI, C.SBORDONE: Γ-convergenza e G-convergenza per problemi non lineari di tipo ellittico. Boll. Un. Mat. Ital. (5) 13 - A (1976), 352 - 362.

[2] L. AMERIO: Continuous solutions of the problem of a string vibrating against an obstacle. Rend. Sem. Mat. Univ. Padova 59 (1978), 67 - 96.

[3] L. AMERIO: Su un problema di vincoli unilaterali per l'equazione non omogenea della corda vibrante. I.A.C., Pubblicazioni, serie D n. 109, 1976, 3 - 11.

[4] L. AMERIO, G. PROUSE: Study of the motion of a string vibrating against an obstacle. Rend. Mat. (6) 8 (1975), 563 - 585.

[5] H. ATTOUCH, C. PICARD: Asymptotic analysis of variational problems with constraints of obstacle type. To appear.

[6] N.S. BAHVALOV, A.A. ZLOTNIK: Coefficient stability of differential equations and averaging equations with random coefficients. Soviet Math. Dokl. 19(1978), 1171 - 1175.

[7] A. BENSOUSSAN, J.L. LIONS, G. PAPANICOLAOU: Asymptotic methods in periodic structures. North Holland, Amsterdam, 1978.

[8] G. BUTTAZZO: Su una definizione generale dei Γ-limiti. Boll. Un. Mat. Ital. (5) 14 - B (1977), 722 - 744.

[9] G. BUTTAZZO, G. DAL MASO: Γ-limits of a sequence of non-convex and non-equi-Lipschitz integral functionals. Ricerche Mat. 27 (1978), 235 - 251.

[10] G. BUTTAZZO, G. DAL MASO: Γ-limits of integral functionals. J. Analyse Math. 37 (1980), 145 - 185.

[11] G. BUTTAZZO, D. PERCIVALE: Sull'approssimazione del problema del rimbalzo unidimensionale. Ricerche Mat., to appear.

[12] G. BUTTAZZO, D. PERCIVALE: On the approximation of the elastic bounce problem on Riemannian manifolds. J. Differential Equations, to appear.

[13] L. CARBONE: Sur la convergence des integrales du type de l'énergie sur des fonctions à gradient borné. J. Math. Pures Appl. (9) 56 (1977), 79 - 84.

[14] L. CARBONE: Sur un problème d'homogéneisation avec des contraintes sur le gradient. J. Math. Pures Appl. (9) 58 (1979), 275 - 297.

[15] L. CARBONE, F. COLOMBINI: On convergence of functionals with unilateral constraints. J. Math. Pures Appl. (9) 59 (1980), 465 - 500.

[16] L. CARBONE, C. SBORDONE: Some properties of Γ-limits of integral functionals. Ann. Mat. Pura Appl. (4) 122 (1979), 1 - 60.

[17] M. CARRIERO, E. PASCALI: Γ -convergenza di integrali non negativi maggiorati da funzionali del tipo dell'area. Ann. Univ. Ferrara Sez. VII 24 (1978), 51 - 64.

[18] M. CARRIERO, E. PASCALI: Il problema del rimbalzo unidimensionale e sue approssimazioni con penalizzazioni non convesse. Rend. Mat. (6) 13 (1980), 541 - 554.

[19] D. CIORANESCU, F. MURAT: Un terme étrange venu d'ailleurs. In "Nonlinear partial differential equations and their applications. Collège de France Seminar. Volume II", ed. by H. Brezis and J.L. Lions. Research Notes in Mathematics 60, Pitman, London, 1982.

[20] C. CITRINI: Controesempi all'unicità del moto di una corda in presenza di una parete. Atti Accad. Naz. Lincei, Rend. Cl. Sci. Fis. Mat. Natur. (8) 67 (1979), 179 - 185.

[21] C. CITRINI: Discontinuous solutions of a nonlinear hyperbolic equation with unilateral constraint. Manuscripta Math. 29 (1979), 323 - 352.

[22] G. DAL MASO: Asymptotic behaviour of minimum problems with bilateral obstacles. Ann. Mat. Pura Appl. (4) 129 (1981), 327 - 366.

[23] G. DAL MASO, P. LONGO: Γ-limits of obstacles. Ann. Mat. Pura Appl. (4) 128 (1980), 1 - 50.

[24] E. DE GIORGI: Sulla convergenza di alcune successioni di integrali del tipo dell'area. Rend. Mat. (6) 8 (1975), 277 - 294.

[25] E. DE GIORGI: Generalized limits in calculus of variations. In "Topics in Functional Analysis 1980 - 81" by F. Strocchi, E. Zarantonello, E. De Giorgi, G. Dal Maso, L. Modica, Scuola Normale Superiore, Pisa, 1981, 117 - 148.

[26] E. DE GIORGI, F. COLOMBINI, L.C. PICCININI: Frontiere orientate di misura minima e questioni collegate. Scuola Normale Superiore, Pisa, 1972.

[27] E. DE GIORGI, G. DAL MASO, P. LONGO: Γ-limiti di ostacoli. Atti Accad. Naz. Lincei, Rend. Cl. Sci. Fis. Mat. Natur. (8) 68 (1980), 481 - 487.

[28] E. DE GIORGI, T. FRANZONI: Su un tipo di convergenza variazionale. Atti Accad. Naz. Lincei, Rend. Cl. Sci. Fis. Mat. Natur. (8) 58 (1975), 842 - 850.

[29] E. DE GIORGI, T. FRANZONI: Su un tipo di convergenza variazionale. Rendiconti del Seminario Matematico di Brescia. 3 (1979), 63 - 101.

[30] E. DE GIORGI, T. FRANZONI: Una presentazione sintetica dei limiti generalizzati. Porthugaliae Math., to appear.

[31] E. DE GIORGI, A. MARINO, M. TOSQUES: Problemi di evoluzione in spazi metrici e curve di massima pendenza. Atti Accad. Naz. Lincei, Rend. Cl. Sci. Fis. Mat. Natur. (8) 68 (1980), 180 - 187.

[32] E. DE GIORGI, A. MARINO, M. TOSQUES: Funzioni (p, q) - convesse, operatori (p, q)-monotoni ed equazioni di evoluzione. Atti Accad. Naz. Lincei, Rend. Cl. Sci. Fis. Mat. Natur., to appear.

[33] E. DE GIORGI, S. SPAGNOLO: Sulla convergenza degli integrali dell'energia per operatori ellittici del 2 ordine. Boll. Un. Mat. Ital. (4) 8 (1973), 391 - 411.

[34] A.M. DYKHNE: Conductivity of a two dimensional two-phase system. Soviet Physics JEPT 32 (1971) , 63 - 65.

[35] S.M. KOZLOV: Averaging of random operators. Math. USSR Sb. 37 (1980), 167 - 180.

[36] J.L. LIONS: Some methods in the mathematical analysis of systems and their control. Science Press Beijing, China, Gordon and Breach, New York, 1981.

[37] P. MARCELLINI, C. SBORDONE: An approach to the asymptotic behaviour of elliptic-parabolic operators. J. Math. Pures Appl. 56 (1977), 157 - 182.

[38] A. MARINO, M. TOSQUES: Curves of maximal slope for a certain class of non - regular functions. Boll. Un. Mat. Ital. (6) 1 - B (1982), 143 - 170.

[39] A. MARINO, M. TOSQUES: Existence and properties of the curves of maximal slope. To appear.

[40] L. MODICA: Omogeinizzazione con coefficienti casuali. In Atti del Convegno "Studio di problemi - limite della analisi funzionale", Bressanone, Sept. 7 - 9, 1981, Pitagora, Bologna, 1982, 155 - 164.

[41] L. MODICA, S. MORTOLA: Un esempio di Γ -convergenza. Boll. Un. Mat. Ital. (5) 14 - B (1977), 285 - 299.

[42] G. MOSCARIELLO: Γ-convergenza negli spazi sequenziali. Rend. Accad. Sci. Fis. Mat. Napoli (4) 43 (1976), 333 - 350.

[43] G.C. PAPANICOLAOU, S.R.S. VARADHAN: Diffusion in regions with many small holes. In "Stochastic Differential Systems, Filtering and Control", Proceedings of the IFIP - WG 7/1 Working Conference, Vilnius, Lithuania, USSR, Aug. 27-Sept. 2, 1978, ed. by B. Grigelionis, Lect. Notes in Control and Information Sciences, 25, Springer, 1980, 190 - 206.

[44] G.C. PAPANICOLAOU, S.R.S. VARADHAN: Boundary values problems with rapidly oscillating random coefficients. In "Random Fields", Esztergom (Hungary), 1979, ed. by J. Fritz, J.L. Lebowitz and R. Szasz, Colloquia Mathematica Societatis Jànos Bolyai, 27, North Holland, Amsterdam, 1981.

[45] L.C. PICCININI: Remarks on a uniqueness problem for dynamical systems. Boll. Un. Mat. Ital. (5) 18 - A (1981), 132 - 137.

[46] E. SANCHEZ - PALENCIA : Non homogeneous media and vibration theory. Lect. Notes in Physics 127, Springer, 1980.

[47] C. SBORDONE: Su alcune applicazioni di un tipo di convergenza variazionale. Ann. Scuola Norm. Sup. Pisa Cl. Sci. (4) 2 (1975), 617 - 638.

[48] M. SCHATZMAN: Problèmes unilatéraux d'evolution du 2ème ordre en temps. These de Doctorat d'Etat, Univ. Pierre et Marie Curie, Paris, Janvier 1979.

[49] M. SCHATZMAN: A hyperbolic problem of second order with unilateral constraints. J. Math. Pures Appl. 73 (1980), 138 - 191

[50] M. SCHATZMAN: Un problème hyperbolique du deuxième ordre avec contrainte unilatérale: le corde vibrante avec obstacle ponctuel. J. Differential Equations 36 (1980), 295 - 334.

[51] S. SPAGNOLO: Sulla convergenza di soluzioni di equazioni paraboliche ed ellittiche. Ann. Scuola Norm. Sup. Pisa Cl. Sci. (3) 22 (1968), 577 - 597.

[52] L. TARTAR: Cours Peccot au Collège de France, Paris, 1977.

[53] V.V. YURINSKIJ: Averaging elliptic equations with random coefficients. Sib. Math. J. 20 (1980), 611 - 623.

[54] V.V. ZHIKOV, S.M. KOZLOV, O.A. OLEINIK, KHA T'EN NGOAN: Averaging ang G-convergence of differential operators. Russian Math. Surveys 34 (1979), 69 - 147.

ENNIO DI GIORGI
Scuola Normale Superiore, Pisa

GIANNI DAL MASO
Istituto di Matematica, Informatica e Sistemistica, Udine

THE APPROXIMATE FIRST-ORDER AND SECOND-ORDER DIRECTIONAL DERIVATIVES FOR A CONVEX FUNCTION

J.-B. HIRIART-URRUTY (*)

KEY WORDS : convex functions, ε-subdifferential, approximate first-order directional derivative, approximate second-order directional derivative, generalized second derivatives.

0 - INTRODUCTION

The introduction of the approximate subdifferential of a convex function has been proved to be useful in convex optimization, from the theoretical viewpoint as well as for the purposes of devising algorithms. Within the context of necessary and sufficient conditions for almost optimality, it was known from the beginning that to claim that a point x_o is an ε-minimum of f is equivalent to declaring that 0 belongs to the ε-subdifferential of f at x_o. From the point of view of minimization procedures, it is widely recognized that ε-subgradient methods can be more usable than methods using exact subgradients. That is mainly due to the fact that it is often easier to have access to an ε-subgradient than to a subgradient. For what concerns the study and the use of ε-subdifferentials, the past sixteen years can roughly be divided into three periods of time :

1965 - 1970 : definition and first properties of the ε-subdifferential, in conjunction with the development of modern convex analysis ;

*1970 - 1979 :*during this period of time, a lot of minimization methods using the ε-subdifferential have been developed (the so-called ε-subgradient methods). All these procedures are *first-order* methods in the sense that the definition of x_{n+1} from x_n requires us to know the ε_n-subdifferential of f at x_n or at least an element of it.

1979 - 1981 : in the past three years, a revival of interest appeared in the study of the approximate first-order directional derivative of a convex function. Such an object, denoted by $f'_\varepsilon(x\ ;\ d)$, enjoys for $\varepsilon > 0$ some noteworthy properties different as for their nature from those of the exact directional derivative $f'(x\ ;\ d)$. To observe that $x \to f'_\varepsilon(x\ ;\ d)$ is locally Lipschitz has been the starting point for

(*) Address : Université Paul Sabatier (TOULOUSE III)
U.E.R. Mathématiques, Informatique, Gestion
118, route de Narbonne
31062 TOULOUSE Cedex, FRANCE

studies on the behaviour of $f'_\varepsilon(x\ ;\ d)$ as a function of x, ε, etc... Moreover, at those points x_0 where it is not differentiable, the function $x \to f'_\varepsilon(x\ ;\ d)$ admits however a directional derivative in all directions δ. This new object, which we denote by $f''_\varepsilon(x_0\ ;\ d,\ \delta)$, plays the role of an approximation of a second-order directional derivative of f at x_0, even if the latter definition does not always make sense. That is precisely what we call the *"approximate second-order directional derivative of f at x_0"*. The aim of this paper is to survey the main new results which recently have been obtained on approximate first-order directional derivative $f'_\varepsilon(x\ ;\ d)$ as well as on the approximate second-order directional derivative $f''_\varepsilon(x\ ;\ d,\ \delta)$.

Before going further, let us recall some known facts on the differentiation of convex functions. The subdifferential ∂f is a substitute for the notion of gradient of f ; it is defined for all nonsmooth convex functions and enjoys calculus rules similar to those known in differential calculus. For second derivatives, things are different and less pleasant ; they also depend on the viewpoint which is considered. When f is defined on $\mathbb{R}$, ∂f is increasing and therefore differentiable almost everywhere (a.e.). If differentiation is looked at in the distributional sense, the second derivative of f is a positive Radon measure and, conversely, every positive Radon measure is the second derivative of a convex function (Schwartz). This kind of result has been generalized by Dudley [13] to the f defined on $\mathbb{R}^n$: a distribution T on $\mathbb{R}^n$ is a convex function if and only if D^2T is a nonnegative (n, n) matrix-valued Radon measure. For more details in that respect, see Dudley's pithy paper and references therein. Nevertheless, results on differentiation in the distributional sense do not help very much when one considers problems of minimization ; pointwise second derivatives are of higher interest. In this regard, the main result goes back to Alexandroff (1939) and is reported as follows : *every convex function is twice differentiable a.e.*. However, this statement should be read cautiously because it does not quite mean what "twice differentiable" usually means in differential calculus ; that will be explained in full details in §II.4. Anyway, the second derivative D^2f of a convex function f can be defined without any ambiguity at almost all points. At the remaining points x_0, under mild assumptions of f, it is possible to define what we call *the second-order directional derivative of* f at x_0 : $d \to f''(x_0\ ;\ d,\ d)$. All the properties of $f''_\varepsilon(x_0\ ;\ d,\ d)$ which we shall survey in the present paper emphasize that this approximate second-order directional derivative indeed plays the role of an approximation of $f''(x_0\ ;\ d,\ d)$.

It remains however that it would be of interest to have the second derivative of f defined as a multifunction $\partial^2 f$ from $\mathbb{R}^n$ into the set of positive semidefinite linear mappings on $\mathbb{R}^n$. So far, no real breakthrough has been made in that direction and the approaches which consist on differentiating the multifunction ∂f have not yielded very significant results.

I - PRELIMINARY DEFINITIONS AND PROPERTIES

Given a lower-semicontinuous convex function f from $\mathbb{R}^n$ into $(-\infty, +\infty]$, the ε-*subdifferential* of f at $x_o \in$ dom f (dom f is the set where f is finite) is defined for each $\varepsilon \geq 0$ as the set of vectors $x^* \in \mathbb{R}^n$ satisfying

$$f(x) \geq f(x_o) + \langle x^*, x-x_o \rangle - \varepsilon \quad \text{for all } x \in \mathbb{R}^n. \tag{1.1}$$

The set of such vectors, denoted by $\partial_\varepsilon f(x_o)$, is a closed convex set which reduces to the subdifferential $\partial f(x_o)$ when $\varepsilon = 0$. We recall that the support function of $\partial_\varepsilon f(x_o)$ is given as

$$d \to f'_\varepsilon(x_o ; d) = \inf_{\lambda>0} \{[f(x_o + \lambda d) - f(x_o) + \varepsilon] \lambda^{-1}\}. \tag{1.2}$$

$\partial_\varepsilon f(x_o)$ is nonempty and compact whenever x_o lies in int (dom f). Since only these points will be of interest in the present review, we claim there is no real loss of generality in assuming that f is finite everywhere. To see that, consider $x_o \in$ int (dom f) and $\bar{\varepsilon} > 0$. There exists a compact neighbourhood K of x_o which is included in int (dom f). We then set

$$k = \max\{ \|x^*\| \mid x^* \in \partial_\varepsilon f(x),\ x \in K,\ 0 \leq \varepsilon \leq \bar{\varepsilon}\} ;$$

$$f_k : x \to f_k(x) = \inf_u \{f(u) + k \|x - u\| \}.$$

f_k is a regularized version of f and it has been proved in [14, Section II] that

- . f_k is a finite convex function which coincides with f on K ;
- . $\partial_\varepsilon f_k(x) = \partial_\varepsilon f(x)$ for all $x \in K$ and $\varepsilon \in [0, \bar{\varepsilon}]$.

So, as soon as x lies in a neighbourhood of x_o and ε in a neighbourhood of 0^+, we can replace f by f_k. Therefore, we henceforth assume that the considered function f is finite everywhere.

I.1. Given x_o, a non-null direction d and $\varepsilon \geq 0$, we denote by $q_{d,\varepsilon}$ the approximate difference quotient

$$q_{d,\varepsilon} : \mathbb{R}_+^* \to \mathbb{R}$$
$$\lambda \to [f(x_o+\lambda d) - f(x_o) + \varepsilon]\lambda^{-1}.$$

When $\varepsilon = 0$, we simply use q_d instead of $q_{d,o}$. $\Lambda_d(\varepsilon)$ (or $\Lambda_{d,\varepsilon}(x_o)$ as the case may be) is the set of $\lambda_o > 0$ for which

$$[f(x_o + \lambda_o d) - f(x_o) + \varepsilon]\,\lambda_o^{-1} = f'_\varepsilon(x_o\,;\,d).$$

In particular, $\Lambda_d(0)$ can be described as $]0, a_d^*]\!]$ (*) where a_d^* is the supremum of all $a \geq 0$ for which

$$f(x_o + \lambda d) = f(x_o) + \lambda f'(x_o\,;\,d) \text{ for all } 0 \leq \lambda \leq a.$$

If $0 < a_d^* < +\infty$, that means that $f_d : \lambda \to f(x_o + \lambda d)$ restricted to the segment $[0, a_d^*]$ is an affine function. $a_d^* = +\infty$ corresponds precisely to the case where f_d is an affine function on $\mathbb{R}_+$. Having $a_d^* = 0$ means that

$$[f(x_o + \lambda d) - f(x_o)]\,\lambda^{-1} > f'(x_o\,;\,d) \text{ for all } \lambda > 0.$$

It turns out to be of interest to look at the $q_{d,\varepsilon}$ as functions of λ^{-1}. More precisely, let $r_{d,\varepsilon}$ be defined (for $\varepsilon \geq 0$) by

$$r_{d,\varepsilon}(\mu) = q_{d,\varepsilon}(\tfrac{1}{\mu}) \text{ for all } \mu > 0. \tag{1.3}$$

We set r_d for $r_{d,o}$. $r_d(\mu)$ can be expressed as $\mu h(\frac{d}{\mu})$ where $h : y \to f(x_o + y) - f(x_o)$. The convex set epih is obtained by translating epi f so that the point $(x_o, f(x_o))$ is moved to $(0, 0)$. Moreover, $\mu h(\frac{d}{\mu}) = (h\mu)(d)$, where by definition $h\mu$ is the convex function whose epigraph is μ epi h. For a fixed d, the function $\mu \to (h\mu)(d)$ is known to be a convex function of $\mu \geq 0$ [30, p. 35]. Thus, r_d is a *convex function* on $\mathbb{R}_+$. This property of the $r_{d,\varepsilon}$ can be derived from a duality result (cf. [24] and [31]) ; it also can be looked at as a mere property of convex functions when they are considered from the projective viewpoint.
Since $r_{d,\varepsilon}(\mu)$, $\varepsilon \geq 0$, converges to $f_\infty(d)$ when $\mu \to 0^+$, we agree on posing

$$r_{d,\varepsilon}(0) = f_\infty(d) \text{ whenever } f_\infty(d) < +\infty.$$

Here f_∞ is what is called the recession function of f (or the asymptotic function of f). Unlike for $q_{d,\varepsilon}$, the minimum value of $r_{d,\varepsilon}$ on $\mathbb{R}_+$ $(= f'_\varepsilon(x_o\,;\,d))$ is always attained

(*) When $\alpha, \beta \in \overline{\mathbb{R}}$, $]\alpha, \beta]\!]$ should always be understood as an intervall of $\mathbb{R}$, i.e. $]\alpha, \beta]\!] = \{\gamma \in \mathbb{R} \mid \alpha < \lambda \leq \beta\}$.

whenever $\varepsilon > 0$. If we denote by $M_d(\varepsilon)$ (or $M_{d,\varepsilon}(x_o)$) the set of $\mu_o \geq 0$ for which $r_{d,\varepsilon}(\mu_o) = f'_\varepsilon(x_o\,;\,d)$, $M_d(\varepsilon)$ is a *nonempty compact interval* of $\mathbb{R}_+$ for all $\varepsilon > 0$. As for $\varepsilon = 0$, $M_d(0)$ can be described as the segment

$$[1/a^*_d,\ +\infty[\quad (1/+\infty = 0 \text{ and } 1/0 = +\infty \text{ by convention}). \tag{1.4}$$

Due to the relationship between $q_{d,\varepsilon}$ and $r_{d,\varepsilon}$, $\mu_o > 0$ is a minimum of $r_{d,\varepsilon}$ on $\mathbb{R}^*_+$ if and only if $\lambda_o = 1/\mu_o$ is a minimum of $q_{d,\varepsilon}$ on $\mathbb{R}^*_+$. Whence

$$\Lambda_d(\varepsilon) = \{\frac{1}{\mu_o} \mid \mu_o \in M_d(\varepsilon),\ \mu_o > 0\} \text{ for all } \varepsilon \geq 0. \tag{1.5}$$

The properties of $\Lambda_d(\varepsilon)$ (or of $M_d(\varepsilon)$) depend on the range of the function $q_{d,\varepsilon}$ as well as on the value taken by $f_\infty(d)$. We recall here the three possibilities to be taken into account :

(S_1) $\Lambda_d(\varepsilon)$ is nonempty and bounded if and only if there is $\lambda_* > 0$ for which $q_{d,\varepsilon}(\lambda^*) < f_\infty(d)$;

(S_2) $\Lambda_d(\varepsilon)$ is nonempty and unbounded if and only if $\text{Min}\{q_{d,\varepsilon}(\lambda) \mid \lambda > 0\} = f_\infty(d)$;

(S_3) $\Lambda_d(\varepsilon)$ is empty if and only if $q_{d,\varepsilon}(\lambda) > f_\infty(d)$ for all $\lambda > 0$.

When f_d is affine on $\mathbb{R}_+$, i.e. $a^*_d = +\infty$, $\Lambda_d(\varepsilon)$ is empty for all $\varepsilon > 0$. If $a^*_d < +\infty$, there necessarily exists $\bar{\varepsilon} > 0$ satisfying :

$$\exists\lambda > 0 \qquad q_{d,\bar{\varepsilon}}(\lambda) < f_\infty(d).$$

Define ε^*_d as the supremum of all the $\bar{\varepsilon} > 0$ for which the above holds. While a^*_d takes into account the behaviour of f_d near 0^+, ε^*_d depends on the behaviour of $f_d(\lambda)$ when $\lambda \to +\infty$. Clearly, $\varepsilon^*_d < +\infty$ if and only if $f_\infty(d) < +\infty$. Due to the definition of ε^*_d, we have the following correspondence between the values of ε and the situations (S_i) :

(i) $0 < \varepsilon < \varepsilon^*_d$: (S_1) ;

(ii) $\varepsilon^*_d = \varepsilon$: (S_2) or (S_3) ;

(iii) $\varepsilon^*_d < \varepsilon$: (S_3).

It is convenient to extend the definition of ε_d^* to the case where f_d is affine on $\mathbb{R}_+$ by posing $\varepsilon_d^* = 0$.

I.2. The main result concerning the behaviour of the function $v_{d,\varepsilon} : x \to f'_\varepsilon(x ; d)$ is that it is *locally Lipschitz* whenever $\varepsilon > 0$ [29, 14]. Thus, generalized gradient techniques [8, 9, 10] can be applied to such functions. At those points x_o where $v_{d,\varepsilon}$ is not differentiable, it admits however a directional derivative $v'_{d,\varepsilon}(x_o ; \delta)$ for all δ. The directional derivative of $v_{d,\varepsilon}$ at x_o in the δ direction is precisely what we define as $f''_\varepsilon(x_o ; d, \delta)$. The exact formulation of $f''_\varepsilon(x_o ; d, \delta)$ was worked out by Lemaréchal and Nurminskii [24] under the assumption that f is coercive (that is $\lim_{\|x\| \to +\infty} f(x)/\|x\| = +\infty$) and generalized by Auslender [3] to arbitrary finite functions. Their result is a follows :

$$f''_\varepsilon(x_o ; d, \delta) = \min_{\mu \in M_d(\varepsilon)} \max_{x^* \in \partial_\varepsilon f(x_o)_d} \{\mu[\langle x^*, \delta\rangle - f'(x_o ; \delta)]\}, \tag{1.6}$$

where

$$\partial_\varepsilon f(x_o)_d = \{x^* \in \partial_\varepsilon f(x_o) \mid \langle x^*, d\rangle = f'_\varepsilon(x_o ; d)\}.$$

When δ equals d, the above formula reduces to

$$f''_\varepsilon(x_o ; d, d) = \bar{\mu}_d(\varepsilon) [f'_\varepsilon(x_o ; d) - f'(x_o ; d)] \tag{1.7}$$

where $\bar{\mu}_d(\varepsilon)$ stands for $\min \{\mu \mid \mu \in M_d(\varepsilon)\}$.

Following the classification we have recalled in the previous paragraph, we retain that

(i) $f''_\varepsilon(x_o ; d, \delta) = 0$ whenever $\varepsilon > \varepsilon_d^*$;

(ii) if $\varepsilon < \varepsilon_d^*$, $f''_\varepsilon(x_o;d,\delta) = [\psi^*_{\partial_\varepsilon f(x_o)_d}(\delta) - f'(x_o ; \delta)] / \lambda_d(\varepsilon)$, where

$\lambda_d(\varepsilon) = \max\{\lambda \mid \lambda \in \Lambda_d(\varepsilon)\}$ or $\min\{\lambda \mid \lambda \in \Lambda_d(\varepsilon)\}$ according as $\psi^*_{\partial_\varepsilon f(x_o)_d}(\delta) - f'(x_o ; \delta)$ is positive or not (*).

To illustrate the foregoing, it is worth-while to consider the one-dimensional case. Given a convex function $f : \mathbb{R} \to \mathbb{R}$, $\partial_\varepsilon f(x)$ is, for all x, a compact interval containing $\partial f(x)$. One can express $\partial_\varepsilon f(x)$ as

$$\partial_\varepsilon f(x) = [\underline{D}_\varepsilon f(x), \bar{D}_\varepsilon f(x)], \tag{1.8}$$

(*) ψ_A^* denotes the support function of A.

where $\underline{D}_\varepsilon f$ and $\overline{D}_\varepsilon f$ are two locally Lipschitz functions on $\mathbb{R}$, namely

$$\overline{D}_\varepsilon f(x) = f'_\varepsilon(x\ ;\ 1),\ \underline{D}_\varepsilon f(x) = -f'_\varepsilon(x\ ;\ -1). \tag{1.9}$$

Let $M^+_\varepsilon(x_o)$ denote $M_{d,\varepsilon}(x_o)$ when the chosen direction d equals 1 ; we set $M^+_\varepsilon(x_o) = [\underline{\mu}^+_\varepsilon(x_o), \overline{\mu}^+_\varepsilon(x_o)]$. The right derivative and the left derivative of $\overline{D}_\varepsilon f$ are given as follows :

$$\forall x_o \in \mathbb{R} \qquad \begin{aligned} (\overline{D}_\varepsilon f)'_+(x_o) &= \underline{\mu}^+_\varepsilon(x_o)\ [\overline{D}_\varepsilon f(x_o) - f'_+(x_o)] \\ (\overline{D}_\varepsilon f)'_-(x_o) &= \overline{\mu}^+_\varepsilon(x_o)\ [\overline{D}_\varepsilon f(x_o) + f'_-(x_o)]. \end{aligned} \tag{1.10}$$

As for the generalized gradient of $\overline{D}_\varepsilon f$, we have that

$$\forall x_o \in \mathbb{R} \qquad \partial(\overline{D}_\varepsilon f)(x_o) = M^+_\varepsilon(x_o)\ [\overline{D}_\varepsilon f(x_o) - \partial f(x_o)]. \tag{1.11}$$

The above is a particular case of the expression of $\partial v_{d,\varepsilon}(x_o)$ when $\partial_\varepsilon f(.)_{\mathbf{d}}$ is single-valued at x_o [16, Corollary 3.7].

Similar formulae can be obtained mutatis mutandis for $\underline{D}_\varepsilon f$.
As an application, the reader is invited to verify (1.10) and (1.11) for $f(x) = |x|$ and $x_o = -\varepsilon/2$.

Before concluding these preliminaries, a word concerning notations. Of course, the above-defined objects $f'_\varepsilon(x_o\ ;\ d)$, $M_{d,\varepsilon}(x_o)$, $f''_\varepsilon(x_o\ ;\ d,\ d)$ depend on the parameters f, ε, x_o and d. To avoid cumbersome notations, all the parameters will not show up in the notations ; it should be clear from the context what parameters are set fixed.

II - PROPERTIES OF THE APPROXIMATE FIRST-ORDER DIRECTIONAL DERIVATIVE

As a function of the variable x, $f'(x\ ;\ d)$ is upper-semicontinuous at all points x_0 and continuous at those points x_0 where f is differentiable. Having available the $f'(x\ ;\ d)$ for all x and d allows us to recover f from them since

$$f(x) = f(x_0) + \int_0^1 f'(x_0 + t(x-x_0)\ ;\ x - x_0)\ dt. \tag{2.1}$$

However, $f'(.;d)$ is differentiable only at some privileged points which we shall consider later.

The perturbed form $f'_\varepsilon(x\ ;\ d)$ enjoys, for $\varepsilon > 0$, some noteworthy properties of regularity, different from those of $f'(x\ ;\ d)$. We shall review them by considering the behaviour of $f'_\varepsilon(x\ ;\ d)$ successively as a function of d, ε, f, x.

II.1. Behaviour of $f'_\varepsilon(x\ ;\ d)$ as a function of d.

Actually there is little to say in regard of the function

$$\begin{aligned} s_\varepsilon\ :\ &\mathbb{R}^n \to \mathbb{R} \\ &d \to s_\varepsilon(d) = f'_\varepsilon(x\ ;\ d) \end{aligned}$$

when f, ε and x are fixed. Since it is the support function of $\partial_\varepsilon f(x)$, s_ε is a positively homogeneous (finite) convex function. The subdifferential of s_ε at d is just the set $\partial_\varepsilon f(x)_d$ involved in the formulation of $f''_\varepsilon(x\ ;\ d,\ \delta)$. The structure of $\partial_\varepsilon f(x)_d$ is made more clear in the next statement (cf. [16] and references therein).

Proposition 2.1.

(i) $\partial_\varepsilon f(x)_d \cap \partial f(x)$ *is empty whenever* $\Lambda_\varepsilon(d)$ *is nonempty ;*

(ii) $\partial_\varepsilon f(x)_d = \partial_\varepsilon f(x) \cap \partial f_\infty(d)$ *if* $f'_\varepsilon(x\ ;\ d) = f_\infty(d)$ *;*

(iii) Assuming $\Lambda_\varepsilon(d)$ *is nonempty, we have that*

$$\partial_\varepsilon f(x\ )_d = \{x^* \in \partial f(x + \lambda d)\ \mid\ \langle x^*, d\rangle = f'_\varepsilon(x\ ;\ d)\}$$

whatever $\lambda \in \Lambda_\varepsilon(d)$.

Since $\partial_\varepsilon f(x)_.$ is a subdifferential multifunction, $\partial_\varepsilon f(x)_d$ is single-valued at almost all d. There are however situations where $\partial_\varepsilon f(x)_d$ is single-valued for all (non-null) d. That obviously holds for all f defined on the real line. Moreover, if f is a differentiable function and if d is a direction for which $\Lambda_\varepsilon(d) \neq \phi$, it comes from the last relation in the proposition above that

$$\partial_\varepsilon f(x)_d = \{\nabla f(x+\lambda d)\} \text{ for any } \lambda \in \Lambda_\varepsilon(d). \quad (2.2)$$

Consequently, we have the following

<u>*Proposition 2.2.*</u> *Assume f is differentiable. Then*

$$f'_\varepsilon(x\,;\,d) = \langle\nabla f(x), d\rangle \text{ if } \Lambda_\varepsilon(d) \text{ is empty } ;$$

$$f'_\varepsilon(x\,;\,d) = \langle\nabla f(x+\lambda d), d\rangle \text{ for any } \lambda \in \Lambda_\varepsilon(d) \text{ when } \Lambda_\varepsilon(d) \text{ is nonempty.}$$

If f is a coercive function on $\mathbb{R}^n$ (i.e. $\lim_{\|x\|\to\infty} f(x)/\|x\| = +\infty$), it is secured that $\Lambda_\varepsilon(d) \neq \phi$ for all x and all d. As a result, the gradient mapping ∇f is a continuous selection of the multifunction $\operatorname{int}(\partial_\varepsilon f)$, that is to say

$$\nabla f(x) \in \operatorname{int}(\partial_\varepsilon f(x)) \text{ for all } x \in \mathbb{R}^n. \quad (2.3)$$

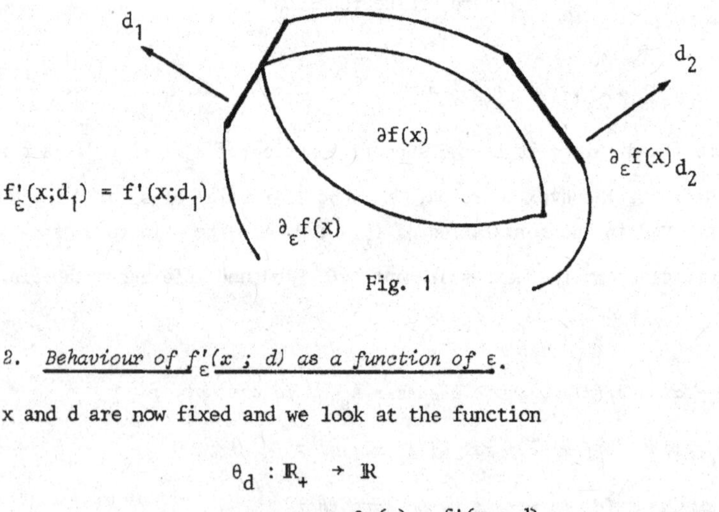

Fig. 1

II.2. <u>*Behaviour of $f'_\varepsilon(x\,;\,d)$ as a function of ε.*</u>

f, x and d are now fixed and we look at the function

$$\theta_d : \mathbb{R}_+ \to \mathbb{R}$$
$$\varepsilon \to \theta_d(\varepsilon) = f'_\varepsilon(x\,;\,d).$$

The qualitative behaviour of θ_d as a function of ε as well as the limiting behaviour of $\theta_d(\varepsilon)$ when $\varepsilon \to 0^+$ have been studied in [17]. We present here the main results in that respect.

First of all, since it is the infimum of a collection of affine functions of ε, θ_d is a *concave* function of the parameter ε. The superdifferential of θ_d at $\varepsilon > 0$ along with the (concave) conjugate function of θ_d are described in the following

Proposition 2.3.

(i) $\partial\theta_d(\varepsilon) = M_d(\varepsilon)$ *for all* $\varepsilon > 0$;

(ii) $r_d(\mu) = \sup_{\varepsilon>0} [f'_\varepsilon(x ; d) - \varepsilon\mu]$ *for all* $\mu \geq 0$.

θ_d is strictly increasing on $[0, \varepsilon^*_d[$ and takes a constant value ($\theta_d(\varepsilon) = f_\infty(d)$) on $]\varepsilon^*_d, +\infty[$. There are two possible situations in regard to the behaviour of $\theta_d(\varepsilon)$ when $\varepsilon \to +\infty$:

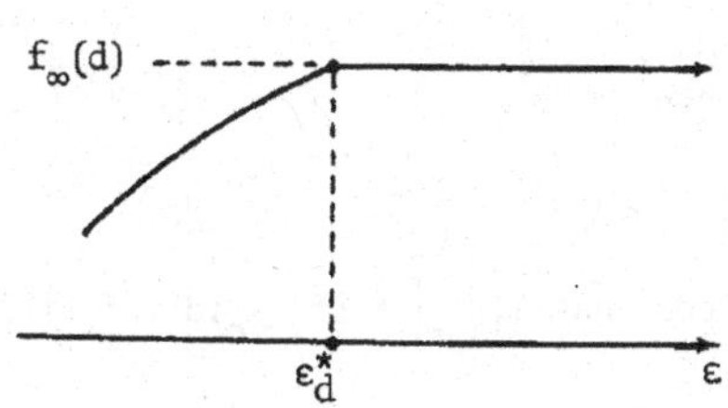

first case : $f_\infty(d) < +\infty$

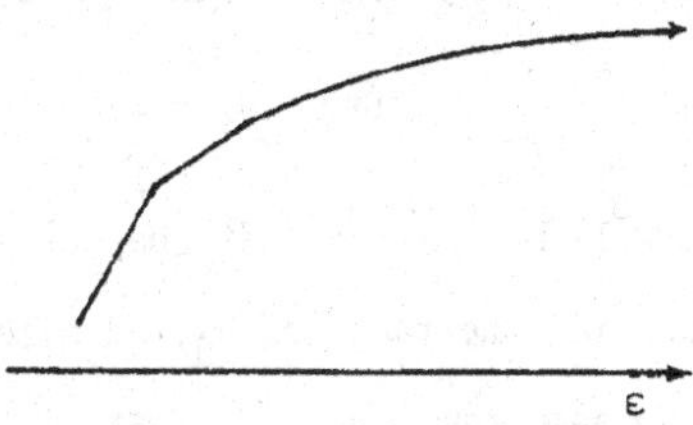

second case : $f_\infty(d) = +\infty$

Fig. 2.

As for the behaviour of $\theta_d(\varepsilon)$ when $\varepsilon \to 0^+$, the important thing to tackle is the right derivative $(\theta_d)'_+(0)$ of θ_d at 0.

Proposition 2.4.

(i) $(\theta_d)'_+(0) = 1/a^*_d$ *in* $\overline{\mathbb{R}}_+$;

(ii) Assuming that $a^*_d < +\infty$, *we have that*

$$\lim_{\varepsilon\to 0^+} [f'_\varepsilon(x_o ; d) - f'(x_o ; d) - \frac{\varepsilon}{\lambda_d(\varepsilon)}] = 0 \text{ whatever } \lambda_d(\varepsilon) \in \Lambda_d(\varepsilon).$$

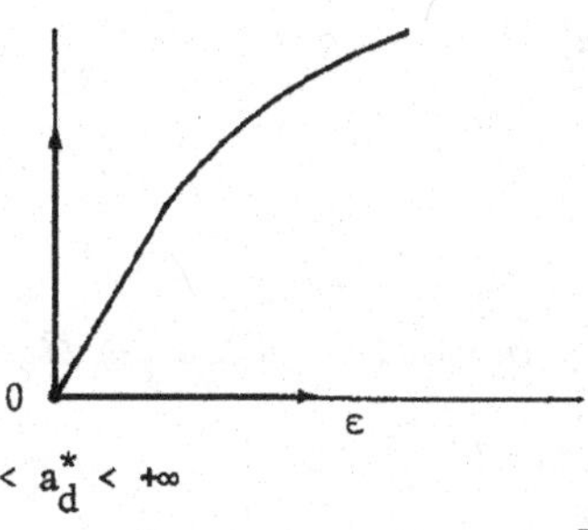

$0 < a^*_d < +\infty$

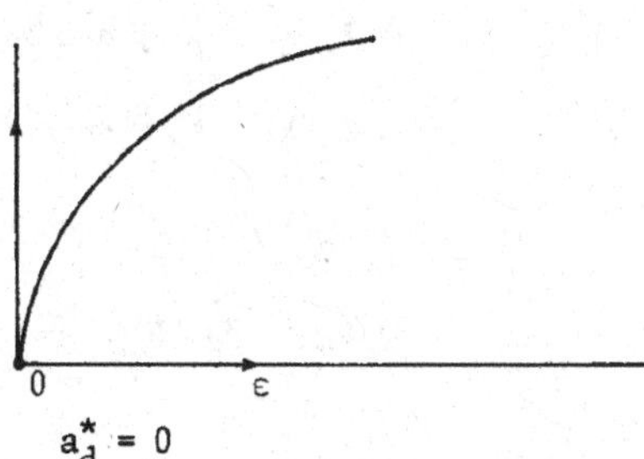

$a^*_d = 0$

Fig. 3.

When $0 < a_d^* < +\infty$, we infer from (ii) in Proposition 2.4 that

$$f'_\varepsilon(x_o \; ; \; d) - f'(x_o \; ; \; d) = \varepsilon/a_d^* + o(\varepsilon) \qquad (2.4)$$

$$= \varepsilon/\Lambda_d(\varepsilon) + o(\varepsilon) \text{ when } \varepsilon \to 0^+. \qquad (2.5)$$

This situation typically occurs when f_d is polyhedral (but with $a_d^* < +\infty$) on $\mathbb{R}_+$. If it is so, the term $o(\varepsilon)$ in (2.4) is null for ε small enough.

When $a_d^* = 0$, an estimate like that of (ii) in Proposition 2.4 is "optimal".As a gener rule, it is hopeless to find $\sigma \in]0, 1[$ for which

$$f'_\varepsilon(x_o \; ; \; d) - f'(x_o \; ; \; d) = O(\varepsilon^\sigma) \text{ when } \varepsilon \to 0^+.$$

More will be said in that respect in Section III.

Concerning the range of θ_d, we already have noticed that $\theta_d(\mathbb{R}_+) = [f'(x_o;d), f_\infty(d)]$. At the same time, we have that

$$\{q_d(\lambda) \mid \lambda > 0\} = [f'(x_o \; ; \; d), f_\infty(d)[.$$

Putting these two facts together merely leads to the following mean value theorem al ready observed by C. Lemaréchal [26].

Proposition 2.5. Assume that $f'(x_o \; ; \; d) < f_\infty(d)$. Then

(i) for all $\lambda > 0$, there exists an unique $\zeta_d(\lambda) \in [0, \varepsilon_d^*[$ for which

$$q_d(\lambda) = \frac{f(x_o + \lambda d) - f(x_o)}{\lambda} = f'_{\zeta_d(\lambda)}(x_o \; ; \; d) \; ; \qquad (2.6)$$

(ii) $\zeta_d : \mathbb{R}_+^* \to [0, \varepsilon_d^*[$ is a continuous mapping enjoying the following properties :

- $\zeta_d(\lambda) = 0$ *if and only if* $\lambda \in]0, a_d^*]$;
- $\zeta_d(\lambda) \to \varepsilon_d^*$ *when* $\lambda \to +\infty$;
- $\zeta_d(\lambda)/\lambda \to 0^+$ *when* $\lambda \to 0^+$;
- $\zeta_d :]a_d^*, +\infty[\to]0, \varepsilon_d^*[$ *is strictly increasing and locally Lipschitz.*

Proof : We have cast aside the degenerate case where $f'(x_o\ ;\ d) = f_\infty(d)$. Indeed, in such a situation, we know that

$$q_d(\lambda) = f'(x_o\ ;\ d) = f'_\zeta(x_o\ ;\ d) \text{ for all } \lambda > 0 \text{ and } \zeta \geq 0.$$

Suppose now that $f'(x_o\ ;\ d) < f_\infty(d)$. On the one hand, we have that

$$\{q_d(\lambda)\ |\ \lambda > 0\} = [f'(x_o\ ;\ d),\ f_\infty(d)[,$$

and $q_d(\lambda)$ equals $f'(x_o\ ;\ d)$ whenever $\lambda \in]0,\ a_d^*]$. On the other hand, θ_d is strictly increasing on $[0,\ \varepsilon_d^*[$ and takes a constant value $(= f_\infty(d))$ on $[\varepsilon_d^*,\ +\infty[$. Whence

$$\{f'_\zeta(x_o\ ;\ d)\ |\ \zeta \in]0,\ \varepsilon_d^*[\} =]f'(x_o\ ;\ d),\ f_\infty(d)[,$$

$$f'_\zeta(x_o\ ;\ d) = f'(x_o\ ;\ d) \text{ if and only if } \zeta = 0.$$

As a result, the unique $\zeta_d(\lambda)$ for which $q_d(\lambda) = f'_{\zeta_d(\lambda)}(x_o\ ;\ d)$ is defined by

$$\zeta_d(\lambda) = 0 \text{ for } \lambda \in]0,\ a_d^*]\ .$$

$$\zeta_d(\lambda) = \theta_d^{-1}[q_d(\lambda)] \text{ for all } \lambda > a_d^*.$$

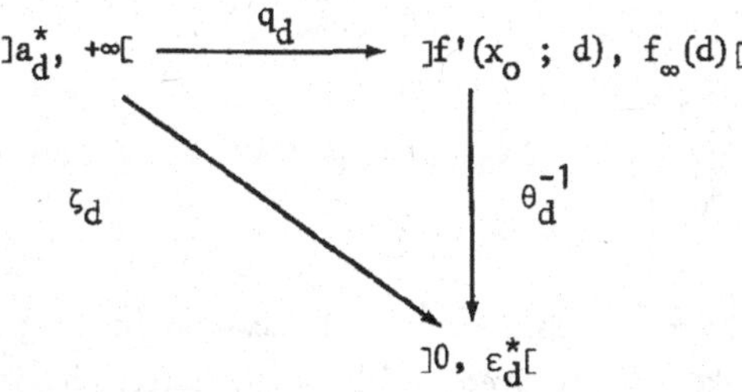

Clearly, the mapping ζ_d inherits its main properties from those of θ_d^{-1} and q_d. The only result which needs some proof concerns the behaviour of $\zeta_d(\lambda)/\lambda$ when $\lambda \to 0^+$, in the case where $a_d^* = 0$. According to Proposition 2.3,

$$q_d(\lambda) = \sup_{\varepsilon>0} [f'_\varepsilon(x_o\ ;\ d) - \frac{\varepsilon}{\lambda}]$$

and the supremum is achieved for those ε for which $\frac{1}{\lambda} \in M_d(\varepsilon)$.

Such ε are obtained by performing the inversion of the multifunction Λ_d,

$$\Lambda_d^{-1} : \mathbb{R}_+^* \rightleftarrows]0, \varepsilon_d^*] .$$

(Here $0 \notin \Lambda_d^{-1}(\lambda)$ for all $\lambda > 0$ because $a_d^* = 0$).
Consequently, by considering $\varepsilon_d(\lambda) \in \Lambda_d^{-1}(\lambda)$ for all λ in a neighbourhood $]0, \bar{\lambda}]$ of 0^+, we have that

$$f'_{\zeta_d(\lambda)}(x_o ; d) = q_d(\lambda) = f'_{\varepsilon_d(\lambda)}(x_o ; d) - \frac{\varepsilon_d(\lambda)}{\lambda} \qquad \forall\lambda \in]0, \bar{\lambda}].$$

It results from this that $\zeta_d(\lambda) < \varepsilon_d(\lambda)$ and, since $\frac{\varepsilon_d(\lambda)}{\lambda} \to 0$ when $\lambda \to 0^+$ (cf. (ii) in Proposition 2.4), we get the desired result. □

Remark 1. Rephrased in terms of r_d instead of q_d, the proposition above states that

$$r_d(\mu) = f'_{\hat{\zeta}_d(\mu)}(x_o ; d) \text{ for all } \mu > 0,$$

where $\hat{\zeta}_d(\mu)$ stands for $\zeta_d(1/\mu)$. Now, since

$$\hat{\zeta}_d = \theta_d^{-1} \circ r_d :]0, 1/a_d^*[\to]0, \varepsilon_d^*[,$$

it is easy to see that $\hat{\zeta}_d$ is *convex* on the interval $]0, 1/a_d^*[$ and that its subdifferential is given as

$$\partial\hat{\zeta}_d(\mu) = \frac{\partial r_d(\mu)}{\partial\theta_d[\hat{\zeta}_d(\mu)]} = \Lambda_d[\hat{\zeta}_d(\mu)]\, \partial r_d(\mu) \text{ for all } \mu \in]0, 1/a_d^*[. \qquad (2.7)$$

Remark 2. A mean value theorem of a different kind shows up in the proof of the proposition above. Let us make it more precise. Given the multifunction

$$\Lambda_d : [0, \varepsilon_d^*] \rightleftarrows \mathbb{R}_+^* ,$$

we denote by E_d the inverse of it, i.e.

$$E_d = \Lambda_d^{-1} : \mathbb{R}_+^* \rightleftarrows [0, \varepsilon_d^*]$$

defined by

$$E_d(\lambda) = \{\varepsilon \geq 0 \mid \lambda \in \Lambda_d(\varepsilon)\} \text{ for all } \lambda > 0.$$

From its definition, we therefore have that

$$q_d(\lambda) = f'_{\varepsilon_d(\lambda)}(x_o\, ;\, d) - \frac{\varepsilon_d(\lambda)}{\lambda} \text{ for all } \lambda > 0 \text{ and } \varepsilon_d(\lambda) \in E_d(\lambda).$$

The properties of E_d can easily be derived from those of Λ_d. Note moreover that $\zeta_d(\lambda) \leq E_d(\lambda)$ for all $\lambda > 0$.

Example : Let f be defined on $\mathbb{R}$ by

$$f(x) = \max\,(x,\, 2x-1,\, 3x-3)$$

and set $x_o = 0$ and $d = 1$. Here $a_d^* = 1$, $\varepsilon_d^* = 3$ and

$$\zeta_d(\lambda) = \begin{cases} 0 \text{ if } \lambda \in]0,\, 1] \\ 1 - 1/\lambda \text{ if } \lambda \in [1,\, 2[\\ 2 - 3/\lambda \text{ if } \lambda \in [2,\, 3[\\ 3 - 6/\lambda \text{ if } \lambda \in [3,\, +\infty[. \end{cases}$$

To illustrate the formula (2.7), let $\mu = 1/3$. We have that

$$\hat{\zeta}_d(\mu) = 1, \quad \partial\theta_d(1) = [1/2,\, 1], \quad \partial r_d(\mu) = \{-3\},$$

while

$$\partial\hat{\zeta}_d(\mu) = [-6,\, -3].$$

II.3. Calculus rules on $f'_\varepsilon(x_o\, ;\, d)$.

Since $d \to f'_\varepsilon(x_o\, ;\, d)$ is the support function of $\partial_\varepsilon f(x_o)$, the calculus rules on the ε-directional derivative are derived from those on the ε-subdifferential and the support functions. For the ε-subdifferential calculus, see [15] and references therein. We sketch here some of the basic calculus rules on the ε-directional derivative.

• *ε-directional derivative of $f_1 + f_2$, of $f \circ A$.* Given two (finite) convex functions f_1 and f_2, we have that

$$(f_1 + f_2)'_\varepsilon\,(x_o\, ;\, d) = \max_{\substack{\varepsilon_1 \geq 0,\ \varepsilon_2 \geq 0 \\ \varepsilon_1 + \varepsilon_2 = \varepsilon}} \{f'_{\varepsilon_1}(x_o\, ;\, d) + f'_{\varepsilon_2}(x_o\, ;\, d)\}.$$

If A is a linear mapping from $\mathbb{R}^m$ to $\mathbb{R}^n$ and f a convex function on $\mathbb{R}^n$, the ε-directional derivative of $f \circ A$ is given as

$$(f \circ A)'_{\varepsilon}(x_o\,;\,d) = f'_{\varepsilon}(Ax_o\,;\,Ad).$$

• *ε-directional derivative of* $\max_{i=1,\ldots,m} f_i$. Let $f_1, \ldots, f_m$ be convex functions and let $f = \max_{i=1,\ldots,m} f_i$. We then have that

$$f'_{\varepsilon}(x_o\,;\,d) = \max\,\{\sum_{i=1}^{m} (\alpha_i f_i)'_{\varepsilon_i}(x_o\,;\,d)\},$$

where the maximum is taken over the α_i and ε_i satisfying

$$\forall i = 1, \ldots, m \qquad \alpha_i \geq 0 \;,\; \varepsilon_i \geq 0$$

$$\sum_{i=1}^{m} \alpha_i = 1 \quad , \quad \sum_{i=1}^{m} \varepsilon_i + f(x_o) - \sum_{i=1}^{m} \alpha_i f_i(x_o) = \varepsilon.$$

The above formula can be generalized to $f = \max_{t \in T} f_t$, where T is a compact index set. This generalization relies on the exact expression of $\partial_{\varepsilon} f$ in terms of $\partial_{\varepsilon_i}(\alpha_i f_{t_i})$; see [32, Theorem 3] for that.

Further calculus rules can be derived from the ε-subdifferential calculus, like the ε-directional derivative of $f_1 \nabla f_2$ (infimal convolution of f_1 and f_2) or that of $g \circ f$ (composed mappings)... Such formulae are useful for deriving calculus rules on the approximate second-order directional derivative (see [18]).

II.4. Behaviour of $f'_{\varepsilon}(x\,;\,d)$ as a function of x.

For fixed $\varepsilon > 0$ and $d \neq 0$, let $v_d : \mathbb{R}^n \to \mathbb{R}$ be defined by

$$\forall x \in \mathbb{R}^n \qquad v_d(x) = f'_{\varepsilon}(x\,;\,d).$$

As indicated in Section I, v_d is locally Lipschitz on $\mathbb{R}^n$ and therefore differentiable almost everywhere. The study of v_d has been carried out in the author's earlier paper [16, §II and III] ; one can find there a classification of all possible situations with regard to the differentiability properties of v_d. Actually, the classification of all possible situations yields three cases, according as $\Lambda_d(x_o)$ is nonempty and bounded, unbounded, or empty. To a large extent it is possible to detect if v_d is differentiable or not at x_o , to decide if $\partial v_d(x_o)$ contains 0 or not, having only $\Lambda_d(x_o)$ at our disposal. The next statements illustrate that possibility.

__Theorem 2.6.__ *The following are equivalent :*

(i) v_d *is differentiable at* x_o *with* $\nabla v_d(x_o) = 0$;

(ii) $\partial v_d(x_o) = \{0\}$;

(iii) $\Lambda_d(x_o)$ *is empty* ;

(iv) $M_d(x_o) = \{0\}$.

As a general rule, M_d is single-valued at x_o (i.e. $M_d(x_o) = \{\mu_d(x_o)\}$) whenever v_d is differentiable at x_o. Moreover, at such a point, we have that

$$\mu_d(x_o)\ \partial_\varepsilon f(x_o)_d = \nabla v_d(x_o) + \mu_d(x_o)\ \partial f(x_o). \tag{2.8}$$

__Theorem 2.7.__ *The following are equivalent :*

(i) v_d *is differentiable at* x_o *with* $\nabla v_d(x_o) \neq 0$;

(ii) $M_d(x_o) = \{\mu_d(x_o)\}, \mu_d(x_o) > 0$ *and* $\partial_\varepsilon f(x_o)_d$ *is a shifted copy of* $\partial f(x_o)$.

Therefore, M_d *is single-valued almost everywhere on* $\mathbb{R}^n$. The same cannot be said for the multifunction $\partial_\varepsilon f(.)_d$; that is due to the fact that $\partial_\varepsilon f(.)_d$ can be set-valued at x_o even when f and v_d are differentiable at x_o (the equality (2.8) is secured because $\mu_d(x_o) = 0$).

__Theorem 2.8.__ *Assume* $\Lambda_d(x_o)$ *is nonempty and bounded and that f is differentiable at* x_o. *Then* v_d *is differentiable at* x_o *if and only if both* M_d *and* $\partial_\varepsilon f(.)_d$ *are single-valued at* x_o.

__Corollary 2.9.__ *Suppose that* $\Lambda_d(x)$ *is nonempty and bounded for almost all* $x \in \mathbb{R}^n$. *Then* $\partial_\varepsilon f(.)_d$ *is single-valued almost everywhere on* $\mathbb{R}^n$.

Note that $\Lambda_d(x)$ is nonempty and bounded for all x whenever $f_\infty(d) = +\infty$.

Theorem 2.10. $\nabla v_d(x)$ *is represented for almost all* x *as*

$$\nabla v_d(x) = \mu_d(x)\,[x_d^*(x) - \nabla f(x)], \tag{2.9}$$

where $x_d^*(x)$ *is the unique element of* $\partial_\varepsilon f(x)_d$ *when* $\nabla v_d(x) \neq 0$ *and any element of* $\partial_\varepsilon f(x)_d$ *whenever* $\nabla v_d(x) = 0$.

The expression of $\nabla v_d(x)$ for almost all x is precisely what is needed for the calculation of the generalized gradient of v_d at any point x_o [16, Proposition 1.11]. Naturally, the set of points where v_d is differentiable depends on the given $\varepsilon > 0$. There is no specific rule governing the way this set varies when ε moves to 0^+. The function obtained at the limit, i.e.

$$w_d : x \to f'(x\,;\,d)$$

does not enjoy any (locally) Lipschitz property. Nevertheless, w_d is differentiable at almost every $x \in \mathbb{R}^n$; let us make it more precise by recalling Alexandroff's and Mignot's statements.

Theorem (Alexandroff, 1939). At almost every $x_o \in \mathbb{R}^n$, *f has a second-order expansion in the sense that there exists a symmetric positive semi-definite* $A^2f(x_o)$ *such that*

$$f(x) = f(x_o) + \langle\nabla f(x_o), x-x_o\rangle + \frac{1}{2}\langle A^2f(x_o)(x-x_o), x-x_o\rangle + o(\|x-x_o\|^2). \tag{2.10}$$

Following Rockafellar [31, p. 887] or Mignot [27, §1.2], the multifunction ∂f is said to be differentiable at x_o if f is differentiable at x_o and if there is a linear mapping denoted by $D^2f(x_o)$ such that

$$\|\partial f(x) - \nabla f(x_o) - D^2f(x_o)(x-x_o)\| = o(\|x-x_o\|), \tag{2.11}$$

or in other words :

$$\forall \eta > 0,\ \exists \delta > 0,\ \forall x \text{ with } \|x-x_o\| \le \delta,\ \forall x^* \in \partial f(x),$$

$$\|x^* - \nabla f(x_o) - D^2f(x_o)(x-x_o)\| \le \eta\|x-x_o\|.$$

$D^2f(x_o)$ will be called the derivative of ∂f at x_o.

Theorem (Mignot, 1976). ∂f is differentiable almost everywhere on $\mathbb{R}^n$.

The above is actually a corollary to Mignot's differentiability theorem on maximal monotone multifunctions [27, Theorem 1.3]. When the multifunction in question is ∂f, more can be said on the operator $D^2 f$. The following was also suggested by Mignot [28].

Proposition 2.11. Let x_o be a point where ∂f is differentiable and let $D^2 f(x_o)$ be its derivative at x_o. Then $D^2 f(x_o)$ is symmetric and positive semi-definite.

Proof. To prove the symmetry of $D^2 f(x_o)$, one proceeds like for twice-differentiable functions. Given s and t in $\mathbb{R}^n$, we define $g : \mathbb{R} \to \mathbb{R}$ by

$$\forall \xi \in \mathbb{R} \qquad g(\xi) = f(x_o + \xi s + t) - f(x_o + \xi s).$$

g is a locally Lipschitz function and, according to chain rules on generalized gradients, we have that

$$\begin{aligned} \partial g(\xi) &\subset \langle \partial f(x_o + \xi s + t), s\rangle - \langle \partial f(x_o + \xi s), s\rangle \\ \partial g(0) &= \langle \partial f(x_o + t), s\rangle - \langle \nabla f(x_o), s\rangle. \end{aligned} \tag{2.12}$$

Let $x_f^*(x_o + t)$ be an arbitrary element of $\partial f(x_o + t)$. By applying the mean value theorem to the function $\xi \to g(\xi) - \xi\langle x_f^*(x_o + t) - \nabla f(x_o), s\rangle$, we get that

$$\begin{aligned} &|g(1) - g(0) - \langle x_f^*(x_o+t) - \nabla f(x_o), s\rangle| \\ &\quad \le \sup_{\substack{\xi \in]0,1[\\ \xi^* \in \partial g(\xi)}} |\xi^* - \langle x_f^*(x_o+t) - \nabla f(x_o), s\rangle|. \end{aligned} \tag{2.13}$$

It comes from the first estimate of (2.12) that

$$\partial g(\xi) \subset \{\langle \partial f(x_o+\xi s+t), s\rangle - \langle \nabla f(x_o), s\rangle\} + \{\langle \nabla f(x_o), s\rangle - \langle \partial f(x_o+\xi s), s\rangle\}.$$

Since ∂f is differentiable at x_o with $D^2 f(x_o)$ as derivative, we have that :

$$\| \langle \partial f(x_o+\xi s+t) - \nabla f(x_o) - D^2 f(x_o).(\xi s+t) \| = o(\|s\| + \|t\|)$$

and

$$\|\nabla f(x_o) - \partial f(x_o+\xi s) - D^2f(x_o).(\xi s)\| = o(\|s\|).$$

Whence

$$|\partial g(\xi) - \langle D^2f(x_o)t,\ s\rangle| = \|s\|\ \ o(\|s\| + \|t\|)$$

and

$$|\langle x_f^*(x_o+t) - \nabla f(x_o),s\rangle - \langle D^2f(x_o)t,s\rangle| = \|s\|\ \ o(\|s\| + \|t\|).$$

Consequently, we infer from (2.13) that

$$|g(1) - g(0) - \langle D^2f(x_o)t,\ s\rangle| = \|s\|\ \ o(\|s\| + \|t\|).$$

Hence, for all $\eta > 0$, there exists δ such that

$$|g(1) - g(0) - \langle D^2f(x_o)t,\ s\rangle| \leq \eta(\|s\| + \|t\|)^2 \tag{2.14}$$

whenever $\|s\| + \|t\| \leq \delta$.

Since $g(1) - g(0) = f(x_o+s+t) - f(x_o+s) - f(x_o+t) + f(x_o)$ is symmetric in s and t, we have the inequality (2.14) with s and t switched. Consequently,

$$|\langle D^2f(x_o)t,\ s\rangle - \langle D^2f(x_o)s,\ t\rangle| \leq 2\eta(\|s\| + \|t\|)^2 \tag{2.15}$$

for $\|s\| + \|t\| \leq \delta$.

Now, since the inequality (2.15) is "homogeneous of degree 2", it holds true for all s and t. Hence the symmetry of $D^2f(x_o)$ is proved.

Let now $s \in \mathbb{R}^n$ and $\lambda > 0$. It comes from the differentiability of ∂f at x_o that

$$\left|\frac{\langle \partial f(x_o+\lambda s) - \nabla f(x_o),\ s\rangle}{\lambda} - \langle D^2f(x_o)s,\ s\rangle\right| = \frac{o(\lambda)}{\lambda}.$$

Given $\eta > 0$, there is $\delta > 0$ such that

$$\forall\lambda \in\ ]0,\delta[\quad \langle D^2f(x_o)s,\ s\rangle \geq \frac{\langle \partial f(x_o+\lambda s) - \nabla f(x_o),\ s\ \rangle}{\lambda} - \eta. \tag{2.16}$$

Since the multifunction ∂f is monotone, $\langle \partial f(x_o+\lambda s) - \nabla f(x_o),\ s\rangle \geq 0$ for all positive λ. Hence, one readily derives from the statement above that $\langle D^2f(x_o)s,\ s\rangle \geq 0$. □

One might wonder whether one can pass from Alexandroff's result to Mignot's one and vice versa. As it is intuitively clear, both statements are equivalent.

Theorem 2.12. *f has a second-order expansion at x_o if and only if ∂f is differentiable at x_o.*

Proof. Suppose that ∂f is differentiable at x_o with $D^2f(x_o)$ as derivative. We consider the function

$$\varphi: x \to \varphi(x) = f(x) - f(x_o) - \langle \nabla f(x_o), x-x_o \rangle - \frac{1}{2} \langle D^2 f(x_o)(x-x_o), x-x_o \rangle.$$

Clearly φ is a locally Lipschitz function satisfying :

$$\varphi(x_o) = 0$$

$$\forall x \in \mathbb{R}^n \qquad \partial\varphi(x) = \partial f(x) - \nabla f(x_o) - D^2 f(x_o)(x-x_o).$$

According to the mean value theorem, there exists $\bar{x} \in]x_o, x[$ and $\bar{x}^* \in \partial\varphi(\bar{x})$ satisfying

$$\varphi(x) = \langle \bar{x}^*, x-x_o \rangle. \tag{2.17}$$

Moreover,

$$\forall \eta > 0, \quad \exists \delta > 0, \quad \forall x \text{ with } \|x - x_o\| \leq \delta, \quad \forall x^* \in \partial\varphi(x),$$

$$\|x^*\| \leq \eta \|x - x_o\|.$$

Consequently, we have that

$$|\varphi(x)| \leq \eta \|x - x_o\|^2$$

whenever $\|x - x_o\| \leq \delta$. Whence $\varphi(x) = o(\|x-x_o\|^2)$.

Conversely, suppose that

$$f(x) = f(x_o) + \langle \nabla f(x_o), x-x_o \rangle + \frac{1}{2} \langle A^2 f(x_o)(x-x_o), x-x_o \rangle + o(\|x-x_o\|^2).$$

We are going to prove that ∂f is differentiable at x_o and that its derivative at x_o is precisely $A^2 f(x_o)$.

Due to the above-mentioned second-order development of f, for all $\eta > 0$ there is $\delta > 0$ such that

$$|f(x) - f(x_o) - \langle\nabla f(x_o),x-x_o\rangle - \frac{1}{2}\langle A^2f(x_o)(x-x_o),x-x_o\rangle| \le \eta\|x-x_o\|^2$$

whenever $\|x-x_o\| \le 2\delta$.

Let Q_η and $\Phi_{\eta,\delta}$ be defined as follows :

$$\forall x \in \mathbb{R}^n \quad Q_\eta(x) = f(x_o) + \langle\nabla f(x_o),x-x_o\rangle + \frac{1}{2}\langle A^2f(x_o)(x-x_o),x-x_o\rangle + \eta\|x-x_o\|^2,$$

$$\Phi_{\eta,\delta}(x) = \begin{cases} Q_\eta(x) \text{ if } \|x-x_o\| \le 2\delta \\ +\infty \text{ if not.} \end{cases}$$

We have that

$$f(x) \le \Phi_{\eta,\delta}(x) \text{ for all } x,$$

$$f(x) \ge Q_\eta(x) - 2\eta\|x-x_o\|^2 \text{ whenever } \|x-x_o\| \le 2\delta.$$

Consider now x satisfying $\|x-x_o\| \le \delta$ and $x^* \in \partial f(x)$. Since

$$f^*(x^*) + f(x) - \langle x, x^*\rangle = 0,$$

we deduce from the preceding inequalities that

$$\Phi^*_{\eta,\delta}(x^*) + Q_\eta(x) - 2\eta\|x-x_o\|^2 - \langle x,x^*\rangle \le 0,$$

which is the same as

$$\Phi^*_{\eta,\delta}(x^*) + \Phi_{\eta,\delta}(x) - \langle x,x^*\rangle \le 2\eta\|x-x_o\|^2,$$

that is

$$x^* \in \partial_\varepsilon \Phi_{\eta,\delta}(x) \text{ where } \varepsilon = 2\eta\|x-x_o\|^2.$$

It comes from chain rules on ε-directional derivatives [15] that :

$$\forall d \in \mathbb{R}^n \qquad (\Phi_{\eta,\delta})'_\varepsilon(x;d) \le (Q_\eta)'_\varepsilon(x;d) + 2\eta\|x-x_o\|.\|d\|$$

$$(Q_\eta)'_\varepsilon(x;d) = \langle A^2f(x_o)(x-x_o), d\rangle + \langle\nabla f(x_o), d\rangle + 2\eta\langle x-x_o, d\rangle$$
$$+ (2\varepsilon[\langle A^2f(x_o)d, d\rangle + 2\eta\|d\|^2])^{1/2}.$$

We therefore get for all d :

$$f'(x;d) - \langle \nabla f(x_o),d\rangle - \langle A^2f(x_o)(x-x_o),d\rangle \leq O(1)\eta\|x-x_o\|.\|d\| + O(1)\sqrt{\eta}\,\|x-x_o\|.\|d\|.$$

Thus $\partial f(x) - \nabla f(x_o) - A^2f(x_o)(x-x_o)$ is contained in the ball of radius $O(1)(\eta + \sqrt{\eta})$ whenever $\|x-x_o\| \leq \delta$. Hence the announced result is proved. □

When f is differentiable in a neighborhood of x_o, to say that ∂f is differentiable at x_o is tantamount to having f twice differentiable at x_o in the usual sense of differential calculus. We thus have :

Corollary 2.13. *Suppose that f is differentiable in a neighborhood of x_o. Then f is twice differentiable at x_o if and only if f has a second-order expansion at x_o.*

So, by extension, we shall say that "f is twice differentiable at x_o" whenever ∂f is differentiable at x_o or, equivalently, f has a second-order expansion at x_o. The linear mapping $D^2f(x_o)$ is called the "second derivative of f at x_o".
Consider a sequence $\{\varepsilon_n\} \subset \mathbb{R}_+^*$ converging to 0. We know that, for all n, v_{d,ε_n} $(x \to f'_{\varepsilon_n}(x\,;\,d))$ is differentiable almost everywhere. Also it is clear from the definitions that the function w_d $(x \to f'(x\,;\,d))$ is differentiable at x_o whenever f is twice differentiable at x_o ; its derivative at x_o then is $D^2f(x_o)\,d$. In addition to that, the sequence of functions $\{v_{d,\varepsilon_n}\}$ converges to w_d pointwise. As shown by the next theorem, at a point x_o belonging to a set of full measure on $\mathbb{R}^n$, the convergence of $\nabla v_{d,\varepsilon_n}(x_o)$ towards $\nabla w_d(x_o)$ is secured.

<u>*Corollary 2.12.*</u> *Let d be a direction for which*

$$\forall a > 0 \quad f(x_o + ad) > f(x_o) + a\, f'(x_o\, ;\, d).$$

Then, for almost all $x_o \in \mathbb{R}^n$*, we have that*

$$\text{(i)} \quad \lim_{n \to +\infty} \nabla v_{d,\varepsilon_n}(x_o) = \nabla w_d(x_o) = D^2 f(x_o)\, d\ ;$$

$$\text{(ii)} \quad \lim_{n \to +\infty} \varepsilon_n / \lambda^2_{d,\varepsilon_n}(x_o) = \frac{1}{2} \langle D^2 f(x_o) d,\, d\rangle\ .$$

Proof. Let E be the set of full measure where f is twice-differentiable and all the v_{d,ε_n} are differentiable. For n large enough, $\Lambda_{d,\varepsilon_n}(x_o)$ is nonempty and contains $\lambda_{d,\varepsilon_n}(x_o)$ as unique element. The announced results then are a particular case of [16, Corollary 4.5]. □

III - PROPERTIES OF THE APPROXIMATE SECOND-ORDER DIRECTIONAL DERIVATIVE

We know from Section I that

(i) $f''_\varepsilon(x_o\, ;\, d,\, \delta) = \min\limits_{\mu \in M_{d,\varepsilon}(x_o)} \{\mu[\psi^*_{\partial_\varepsilon f(x_o)_d}(\delta) - f'(x_o\, ;\, \delta)]\}$,

(ii) $f''_\varepsilon(x_o\, ;\, d,\, d) = \bar{\mu}_{d,\varepsilon}(x_o)\, [f'_\varepsilon(x_o;d) - f'(x_o;d)]$ where $\bar{\mu}_{d,\varepsilon}(x_o)$ stands for $\min\{\mu \mid \mu \in M_{d,\varepsilon}(x_o)\}$.

It is clear that the behaviour of $f''_\varepsilon(x\, ;\, d,\, d)$ as a function of x, ε, d depends on the properties of $f'_\varepsilon(x\, ;\, d)$ as a function of x, ε, d (what is known from Section II) but also on the variations of $M_{d,\varepsilon}(x)$ as a multifunction of x, ε, d.

III.1. *Semicontinuity of $f''_\varepsilon(x\ ;\ d,\ d)$ as a function of x.*

The definition itself of $M_{d,\varepsilon}(x)$ and the continuity properties of f make that the multifunction $x \rightrightarrows M_{d,\varepsilon}(x)$ is upper-semicontinuous at any x_o and bounded in a neighbourhood of x_o. Therefore, the mapping $x \to \bar{\mu}_{d,\varepsilon}(x)$ is lower-semicontinuous and, thanks to the upper-semicontinuity of $x \to f'(x\ ;\ d)$, we get that the function $\psi_{d,\varepsilon} : x \to f''_\varepsilon(x;d,d)$ is *lower-semicontinuous* on $\mathbb{R}^n$. Naturally, if $M_{d,\varepsilon}$ is single-valued at x_o and if f is differentiable at x_o, the function $\psi_{d,\varepsilon}$ is continuous at x_o.

III.2. *Behaviour of $f''_\varepsilon(x_o\ ;\ d,\ d)$ as a function of ε.*

This behaviour has been studied in the author's earlier paper [17, §III] ; we sketch here the main results in that respect. Let ρ_d denote the function $\varepsilon \to f''_\varepsilon(x_o\ ;\ d,\ d)$. Since $\rho_d(\varepsilon) = 0$ for all $\varepsilon \geq \varepsilon^*_d$, the first problem to look at is the *qualitative* behaviour of ρ_d on $]0,\ \varepsilon^*_d[$. ρ_d is actually the quotient of two decreasing functions of ε, namely

$$\forall\varepsilon \in\ ]0,\ \varepsilon^*_d[\qquad \rho_d(\varepsilon) = \frac{f'_\varepsilon(x_o\ ;\ d) - f'(x_o\ ;\ d)}{\bar{\lambda}_d(\varepsilon)},$$

where $\bar{\lambda}_d(\varepsilon)$ stands for $\max\{\lambda \mid \lambda \in \Lambda_d(\varepsilon)\}$. It is thus differentiable almost everywhere and, at a point ε where it exists, the derivative $\rho'_d(\varepsilon)$ can be expressed in terms of $\rho_d(\varepsilon)$ and the derivative $\lambda'_d(\varepsilon)$ of the multifunction Λ_d at ε [17, Theorem 3.2]. This is of course a partial information and additional assumptions have to be made on the behaviour of f on the half-line $x_o + \mathbb{R}_+ d$ to secure, for example, that ρ_d is continuous or locally Lipschitz on $]0,\ \varepsilon^*_d[$ [17, Theorem 3.3].

The *limiting* behaviour of $\rho_d(\varepsilon)$ when $\varepsilon \to 0^+$ is the second question to be considered. Under mild assumptions on the behaviour of f on $x_o + \mathbb{R}_+ d$, $\lim_{\varepsilon \to 0^+} \rho_d(\varepsilon)$ does exist and coincides with what was expected, namely "the second-order directional derivative of f at x_o in the direction d". For that, recall that f has a *second-order Dini derivative* at x_o in the d direction if

$$D''f(x_o)(d) = \lim_{\lambda \to 0^+} \frac{1}{\lambda} [f'(x_o+\lambda d\ ;\ d) - f'(x_o;d)] \tag{3.1}$$

exists in $\mathbb{R}_+$. Likewise, f is said to have a *second-order de la Vallée-Poussin dérivative* at x_o in the d direction if

$$V'' f(x_o)(d) = \lim_{\lambda \to 0^+} \frac{1}{\lambda} \left[\frac{f(x_o+\lambda d) - f(x_o)}{\lambda} - f'(x_o ; d)\right] \tag{3.2}$$

exists in $\mathbb{R}_+$. The existence of $D''f(x_o)(d)$ implies that of $V''f(x_o)(d)$ with $D''f(x_o)(d) = 2V''f(x_o)(d)$. That actually holds true for any locally Lipschitz f admitting directional derivatives ; for a proof see [4, Proposition 2.3] for example. However, for convex functions f, the existence of $D''f(x_o)(d)$ and that of $V''f(x_o)(d)$ are equivalent. According to Alexandroff [1, p.6] or Busemann [7, p. 10], this result dates back to Jessen [19]. So, if one of the above-mentioned limits exists, we shall simply say that f has *a second-order derivative at x_o in the d direction.*

The main result on the behaviour of $f''_\varepsilon(x_o ; d, d)$ when $\varepsilon \to 0^+$ is as follows [17, §III] :

Theorem 3.1. Assume that f has a second-order derivative at x_o in the direction d. Then

$$f''_\varepsilon(x_o ; d, d) \to D''f(x_o)(d) \text{ when } \varepsilon \to 0^+. \tag{3.3}$$

As for an example, consider a polyhedral function f. Given x_o and d, there are two possibles situations : $a^*_d = +\infty$ or $0 < a^*_d < +\infty$. If $a^*_d = +\infty$, there is nothing to say since

$$\forall \varepsilon > 0 \qquad f''_\varepsilon(x_o ; d, d) = D''f(x_o)(d) = 0. \tag{3.4}$$

If $0 < a^*_d < +\infty$, $f''_\varepsilon(x_o ; d, d)$ is linear in ε for ε small enough. Indeed, there is a threshold $\underline{\varepsilon}_d > 0$ such that

$$\forall\ 0 < \varepsilon < \underline{\varepsilon}_d \qquad f'_\varepsilon(x_o ; d) = f'(x_o ; d) + \frac{\varepsilon}{a^*_d}, \qquad \Lambda_d(\varepsilon) = \{a^*_d\}. \tag{3.5}$$

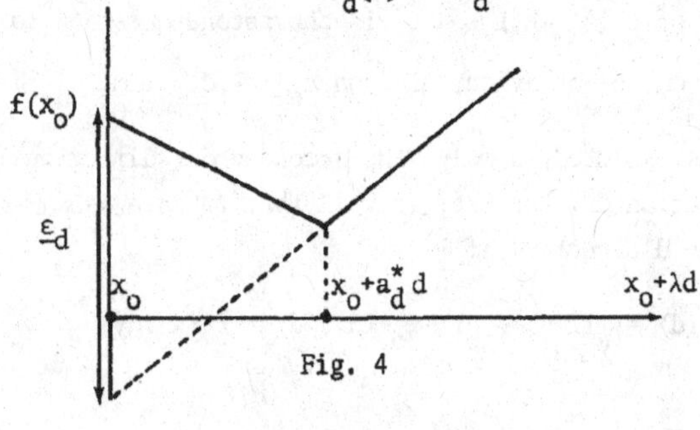

Fig. 4

Thus,

$$f''_\varepsilon(x_o\,;d,\,d) \;=\; \varepsilon/(a^*_d)^2 \text{ for } \varepsilon \text{ small enough.}$$

So, we are in the presence of a *nonsmooth* convex function f for which

$$D''f(x_o)(d) = 0 \text{ for all } x_o \text{ and } d. \tag{3.6}$$

To conclude this paragraph, we pose a problem whose complete solution is unknown to us : *let f be a convex function such that $D''f(x_o)(d)$ exists and is null for all x_o and d ; what can be said about f ?*

III.3. Calculus rules on $f''_\varepsilon(x_o\,; d,\,\delta)$.

Like for the ε-directional derivative f'_ε, calculus rules on the approximate second-order directional derivative are of importance. It is of interest to have expressions of $(f_1+f_2)''_\varepsilon$, $(f\circ A)''_\varepsilon$ $(\max_{i=1,\dots,m} f_i)''_\varepsilon$, etc... in terms of $(f_i)''_{\varepsilon_i}$. The main task for obtaining such calculus rules is to exhibit exact expressions of $M^{f_1+f_2}_{d,\varepsilon}(x_o)$, $M^{f\circ A}_{d,\varepsilon}(x_o)$, etc... in terms of $M^{f_i}_{d_i,\varepsilon_i}(x_i)$. That will be done in a subsequent paper of the author [18]. We just give here an example of chain rule. Given a convex function $f : \mathbb{R}^m \to \mathbb{R}$ and a linear mapping $A : \mathbb{R}^n \to \mathbb{R}^m$, we have that

$$(f\circ A)'_\varepsilon\,(x_o\;;\,d) \;=\; f'_\varepsilon(Ax_o\;;\,Ad)$$

$$M^{f\circ A}_{d,\varepsilon}(x_o) \;=\; M^f_{Ad,\varepsilon}(Ax_o)$$

for all x_o, d, ε > 0. Whence the following chain rule :

$$(f\circ A)''_\varepsilon\,(x_o\,;d,\,d) \;=\; f''_\varepsilon(Ax_o\;;\,Ad,\,Ad).$$

III.4. Behaviour of $f''_\varepsilon(x_o\,; d,\,\delta)$ as a function of $(d,\,\delta)$.

Here again, the behaviour of $M_{d,\varepsilon}(x_o)$ as a multifunction of d is the key-point. Auslender noticed in [3, §2] that

$$\forall\alpha > 0 \qquad M_{\alpha d,\varepsilon}(x_o) \;=\; \alpha\, M_{d,\varepsilon}(x_o). \tag{3.7}$$

However, to prove this equality, he had to consider two cases according as $0 \in M_{d,\varepsilon}(x_o)$ or not. Things are simpler once one has observed that

$$\partial\theta_{d,x_o}(\varepsilon) = M_{d,\varepsilon}(x_o) \text{ for all } \varepsilon > 0,$$

where $\theta_{d,x_o} : \varepsilon \to f'_\varepsilon(x ; d)$ (cf. Proposition 2.3). Clearly, $\theta_{\alpha d,x_o} = \alpha\, \theta_{d,x_o}$ for all $\alpha > 0$ and equality (3.7) readily follows. As a result, we have the following [3, Proposition 2.2] :

$$\forall \alpha, \beta > 0 \quad f''_\varepsilon(x_o ; \alpha d, \beta\delta) = f''_\varepsilon(x_o;\beta d,\alpha\delta) = \alpha\beta\, f''_\varepsilon(x_o ; d, \delta). \tag{3.8}$$

As noticed in [16, §IV.2], the following general inequality holds true

$$f''_\varepsilon(x_o ; d, d) \geq \varepsilon[\bar{\mu}_{d,\varepsilon}(x_o)]^2. \tag{3.9}$$

Hence, in view of (3.8), we rewrite the above as

$$f''_\varepsilon(x_o ; d, d) \geq \varepsilon[\bar{\mu}_{u,\varepsilon}(x_o)]^2.\|d\|^2 \text{ where } u = \frac{d}{\|d\|}. \tag{3.10}$$

Consequently, $f''_\varepsilon(x_o ; d, d)$ goes to $+\infty$ with $\|d\|$ if $\bar{\mu}_{u,\varepsilon}(x_o)$ is kept away from 0. This is certainly true for ε small enough, except for the particular case where f is affine on $x_o + \mathbb{R}_+ d$.

<u>*Proposition 3.2.*</u> *Let $u = d / \|d\|$. We then have that*

$$\lim_{\|d\| \to +\infty} f''_\varepsilon(x_o ; d, d) = +\infty \text{ for all } 0 < \varepsilon < \varepsilon^*_u. \tag{3.11}$$

Proof. For the convenience of the reader, we recall that ε^*_u equals 0 when f is affine on $x_o + \mathbb{R}_+ d$. In such a case, $f''_\varepsilon(x_o ; d, d) = 0$ for all $\varepsilon > 0$. Except in this particular situation, $\varepsilon^*_u > 0$ and

$$M_{u,\varepsilon}(x_o) \subset \mathbb{R}^*_+ \text{ for all } 0 < \varepsilon < \varepsilon^*_u \quad \text{(cf. Section I)}.$$

Hence the result (3.11) is derived from (3.10). □

Remark. Assuming that f is strongly convex on $x_o + \mathbb{R}_+ d$ with ρ as modulus of strong convexity, the following lower bound has been obtained in [3, Proposition 2.3] :

$$\forall d \in \mathbb{R}^n \qquad f''_\varepsilon(x_o\ ;\ d,\ d) \geq 2\sqrt{\varepsilon\rho}\ \ \mu_{u,\varepsilon}(x_o).\|d\|^2,$$

where $u = d/\|d\|$ and $\mu_{u,\varepsilon}(x_o)$ is the unique element of $M_{u,\varepsilon}(x_o)$.

Proposition 3.3. Suppose that $f_\infty(d) < +\infty$ and set $\sigma = f_\infty(u) - f'(x_o\ ;u)$ for $u = d/\|d\|$. Then

$$\forall d \in \mathbb{R}^n \qquad f''_\varepsilon(x_o\ ;\ d,\ d) \leq \frac{\sigma^2}{\varepsilon}\ \|d\|^2. \tag{3.12}$$

Proof. Let $\mu > 0$ be an element of $M_{d,\varepsilon}(x_o)$. According to the definition itself, we have that

$$\mu[f(x_o + \frac{d}{\mu}) - f(x_o)] + \varepsilon\mu = f'_\varepsilon(x_o\ ;\ d).$$

Thus

$$f'(x_o\ ;\ d) + \varepsilon\mu \leq f'_\varepsilon(x_o\ ;\ d) \leq f_\infty(d)$$

$$\Longrightarrow \quad \mu \leq \frac{\sigma}{\varepsilon}\ \|d\|.$$

Whence

$$f''_\varepsilon(x_o\ ;\ d,\ d) = \bar{\mu}_{d,\varepsilon}(x_o)\ [f'_\varepsilon(x_o\ ;\ d) - f'(x_o\ ;\ d)]$$

$$\leq \frac{\sigma}{\varepsilon}\ \|d\|\ (\sigma\ \|d\|). \quad \square$$

It results from (3.8) that $D''_\varepsilon f(x_o)\ :\ d \to f''_\varepsilon(x_o\ ;\ d,\ d)$ is quadratic on each half-line issued from the origin, namely

$$D''_\varepsilon f(x_o)(\alpha u) = \alpha^2\ D''_\varepsilon f(x_o)\,u \text{ for all } \alpha > 0 \text{ and all } u \text{ satisfying } \|u\| = 1.$$

Whence *$D''_\varepsilon f(x_o)$ is convex on each line passing through the origin.*

The question of the convexity of $D''_\varepsilon f(x_o)$ was posed from the beginning (Lemaréchal, 1980) but it rapidly came out that $D^2_\varepsilon f(x_o)$ was not convex even for functions as simple as polyhedral ones [26].

Example (Lemaréchal). Let $f : \mathbb{R}^2 \to \mathbb{R}$ be defined by

$$\forall x = (\xi_1, \xi_2) \quad f(x) = \max(\xi_1, \xi_2, -1 + \xi_1 + \xi_2).$$

We set $x_o = (0, 0)$, $d_\alpha = (\alpha, 1-\alpha)$ for $\alpha \in [0, 1]$. For $0 < \varepsilon < 1$ we have that

$$\Lambda_{d_\alpha,\varepsilon} = \phi \text{ if } \alpha = 0 \text{ or } 1,$$

$$\Lambda_{d_\alpha,\varepsilon} = \{\frac{1}{\min(\alpha,1-\alpha)}\} \text{ if } \alpha \in]0, 1[.$$

Hence $f''_\varepsilon(x_o ; d_\alpha, d_\alpha) = \varepsilon[\min(\alpha, 1-\alpha)]^2$ for all $\alpha \in [0, 1]$. By writing $d_\alpha = \alpha d_1 + (1-\alpha)\, d_o$, one finds out

$$f''_\varepsilon(x_o ; d_o, d_o) = f''_\varepsilon(x_o ; d_1, d_1) = 0$$

$$f''_\varepsilon(x_o ; d_\alpha, d_\alpha) > 0 \text{ for } \alpha \in]0, 1[.$$

In this example $a^*_{d_o} = a^*_{d_1} = +\infty$ while $a^*_{d_\alpha} < +\infty$ for $\alpha \in]0, 1[$. This discrepancy cannot occur when f is differentiable at x_o as the next lemma shows it.

Lemma 3.4. Let f be differentiable at x_o , let d_o and d_1 be two non-null directions and let $\alpha \in [0, 1]$. Then

$$a^*_{\alpha d_1+(1-\alpha)d_o} \geq \min \{a^*_{d_1}, a^*_{d_o}\}. \tag{3.13}$$

Proof. Let $a < \min \{a^*_{d_o}, a^*_{d_1}\}$. By the definition of $a^*_{d_i}$ we have that

$$\forall\lambda \in [0, a] \qquad f(x_o + \lambda d_o) = f(x_o) + \lambda\langle\nabla f(x_o), d_o\rangle$$

$$f(x_o + \lambda d_1) = f(x_o) + \lambda\langle\nabla f(x_o), d_1\rangle.$$

Therefore

$$f[x_o + \lambda(\alpha d_1 + (1-\alpha)d_o)] = f(x_o) + \lambda\langle\nabla f(x_o), \alpha d_1 + (1-\alpha)d_o\rangle$$

for all $\lambda \in [0, a]$.

Whence $a \leq a^*_{\alpha d_1+(1-\alpha)d_o}$ and the inequality (3.13) follows. □

We have recalled in §III.2 that, for a polyhedral f,

$$f''_\varepsilon(x_o\ ;\ d,\ d) = \frac{\varepsilon}{(a^*_d)^2} \text{ for } \varepsilon \text{ small enough.}$$

This result combined with that of the preceding lemma yield the following :

<u>Proposition 3.5.</u> Let f be a polyhedral function differentiable at x_o. There then exists $\underline{\varepsilon} > 0$ such that

$$D''_\varepsilon f(x_o) \text{ is convex for all } 0 < \varepsilon < \underline{\varepsilon}.$$

Proof. For polyhedral f we have that

$$f''_\varepsilon(x_o\ ;\ d,\ d) = \frac{\varepsilon}{(a^*_d)^2} \text{ for } 0 < \varepsilon < \underline{\varepsilon}_d. \tag{3.14}$$

It is not hard to check that there is $\underline{\varepsilon} > 0$ such that $\underline{\varepsilon}_d > \underline{\varepsilon}$ for all d ; this is due to the definition of the threshold $\underline{\varepsilon}_d$ (see §III.2) and to the special structure of f. Combining (3.13) et (3.14) we get that

$$\forall\alpha \in [0,\ 1] \quad D''_\varepsilon f(x_o)(\alpha d_1 + (1-\alpha)d_o) \leq \max\{D''_\varepsilon f(x_o)(d_1),\ D''_\varepsilon f(x_o)(d_o)\}.$$

Moreover, using the same arguments as in §III.1, we observe that $D''_\varepsilon f(x_o)$ is lower-semicontinuous. Thus the function which faces us is quasi-convex, lower-semicontinuous, positive, and positively homogeneous of degree 2 ; such a function necessarily is convex [11, p. 117]. □

Remark. Lemaréchal [25] proved the result of the proposition above by using a different way ; given a polyhedral f

$$f : x \to \max_{i=1,\dots,k} \{\langle a_i,\ x\rangle + b_i\},$$

he exhibited the explicit formulation of $f''_\varepsilon(x_o\ ;\ d,\ d)$ and verified that $D''_\varepsilon f(x_o)$ is convex.

So, as a general rule, $D''_\varepsilon f(x_0)$ is not a convex function. The question of convexity or non-convexity may also be posed for the function $d \to D''f(x_0)(d)$ whenever $D''f(x_0)(d)$ is defined for all d. At a point where f is twice differentiable, we know that

$$D''f(x_0)(d) = \langle D^2 f(x_0)d, d\rangle \text{ for all d.}$$

Whence $D''f(x_0)$ is convex for almost all x_0. A slightly stronger result is as follows :

<u>*Proposition 3.6.*</u> *Assume f is differentiable at x_0 and that $D''f(x_0)(d)$ exists for all d. Then the function $D''f(x_0)$ is convex.*

Proof. The result is immediate once one has written

$$D''f(x_0)(d) = \lim_{\lambda \to 0^+} \frac{1}{\lambda}\left[\frac{f(x_0+\lambda d) - f(x_0)}{\lambda} - \langle \nabla f(x_0), d\rangle\right]. \qquad \square$$

For polyhedral functions f, $D''f(x_0)$ is convex (since identically null !) whether f is differentiable at x_0 or not. So, the question of convexity or non-convexity of $D''f(x_0)$ is not fully answered.

As it is clear from the present study, $f''_\varepsilon(x_0\ ;\ d, d)$ plays the role of an approximation of $D''f(x_0)(d)$ even if the latter concept does not always exist. f''_ε could therefore serve as a substitute for the second-order derivative in devising second-order minimization procedures. For that purposes, more should be known on the behaviour of $f''_\varepsilon(x_0\ ;\ d, d)$ as a function of d ; in that respect, only the first fruits have been mentioned here.

REFERENCES

[1] A.D. ALEXANDROV. *The existence almost everywhere of the second differential of a convex function and some associated properties of convex surfaces* (in Russian), Učenye Zapiski Leningr. Gos. Univ. Ser. Mat. 37 n°6 (1939), 3-35.

[2] E. ASPLUND and R.T. ROCKAFELLAR, *Gradients of convex functions*, Trans. Amer. Math. Soc. 139 (1969), 443-467.

[3] A. AUSLENDER, *Differential properties of the support function of the ε-subdifferential of a convex function*, Note aux Comptes Rendus Acad. Sc. Paris, t. 292 (1981), 221-224 & Math. Programming, to appear.

[4] A. AUSLENDER, *Stability in mathematical programming with nondifferentiable data ; second-order directional derivative for lower-C^2 functions*. Preprint 1981.

[5] M.L. BALINSKI and P. WOLFE, editors, Nondifferentiable Optimization, Math. Programming Study 3, North-Holland (1975).

[6] A. BRØNDSTED and R.T. ROCKAFELLAR, *On the subdifferentiability of convex functions*, Proc. Amer. Math. Soc. 16 (1965), 605-611.

[7] H. BUSEMANN, Convex Surfaces, Interscience Tracts in Pure and Applied Mathematics, 1958.

[8] F.H. CLARKE, *Generalized gradients and applications*, Trans. Amer. Math. Soc. 205, (1975), 247-262.

[9] F.H. CLARKE, *Generalized gradients of Lipschitz functionals*, Advances in Mathematics 40, (1981), 52-67.

[10] F.H. CLARKE, Nonsmooth Analysis and Optimization, John Wiley & Sons, book to appear in 1983.

[11] J.-P. CROUZEIX, Contributions à l'étude des fonctions quasiconvexes, Thèse de Doctorat Es Sciences Mathématiques, Université de Clermont-Ferrand II, (1977).

[12] V.F. DEM'YANOV and V.N. MALOZEMOV, Introduction to Minimax, John Wiley & Sons, 1974.

[13] R.M. DUDLEY, *On second derivatives of convex functions*, Math. Scand. 41 (1977), 159-174 & 46 (1980), 61.

[14] J.-B. HIRIART-URRUTY, *Lipschitz r-continuity of the approximate subdifferential of a convex function*, Math. Scand. 47 (1980), 123-134.

[15] J.-B. HIRIART-URRUTY, *ε-subdifferential calculus*, in Proceedings of the Colloquium "Convex Analysis and Optimization", Imperial College, London (28-29 February 1980), to appear in 1982.

[16] J.-B. HIRIART-URRUTY, *Approximating a second-order directional derivative for nonsmooth convex functions*, SIAM J. on Control and Optimization, to appear in 1982

[17] J.-B. HIRIART-URRUTY, *Limiting behaviour of the approximate first-order and second-order directional derivatives for a convex function*, Nonlinear Analysis : Theory, Methods & Applications, to appear in 1982.

[18] J.-B. HIRIART-URRUTY, *Calculus rules on the approximate second-order directional derivative of a convex function*, in preparation.

[19] B. JESSEN, *Om konvekse Kurvers Krumning*, Mat. Tidsskr. B (1929), 50-62.

[20] S.S. KUTATELADZE, *Convex ε-programming*. Soviet Math. Dokl. 20 (1979), 391-393.

[21] S.S. KUTATELADZE, *ε-subdifferentials and ε-optimization* (in Russian), Sibirskii Matematicheskii Journal (1980), 120-130.

[22] C. LEMARECHAL and R. MIFFLIN, editors, Nonsmooth Optimization, I.I.A.S.A. Proceedings Series, Pergamon Press (1978).

[23] C. LEMARECHAL, Extensions Diverses des Méthodes de Gradient et Applications, Thèse de Doctorat Es Sciences Mathématiques, Paris (1980).

[24] C. LEMARECHAL and E.A. NURMINSKII, *Sur la différentiabilité de la fonction d'appui du sous-différentiel approché*, Note aux Comptes Rendus Acad. Sc. Paris, t. 290 (1980), 855-858.

[25] C. LEMARECHAL, *Some remarks on second-order methods for convex optimization*, *Meetin* "Optimization : Theory & Algorithms" Confolant, 16-20 March 1981.

[26] C. LEMARECHAL, personal communication (March 1981).

[27] F. MIGNOT, *Contrôle dans les inéquations variationnelles elliptiques*, J. of Functional Analysis, Vol. 22 (1976), 130-185.

[28] F. MIGNOT, personal communication (February 1981).

[29] E. A. NURMINSKII, *on ε-differential mapping and their applications in nondifferentiable optimization*, Working paper 78-58, I.I.A.S.A., December 1978.

[30] R.-T. ROCKAFELLAR, Convex Analysis, Princeton University Press, 1970.

[31] R.-T. ROCKAFELLAR, *Monotone operators and the proximal point algorithm*, SIAM J. Control & Optimization 14 (1976), 877-898.

[32] J.-J. STRODIOT, NGUYEN VAN HIEN and N. HEUKEMES, *ε-optimal solutions in nondifferentiable convex programming and some related questions*. Department of Mathematics, University of Namur, preprint 1980.

New Applications of Nonsmooth Analysis to Nonsmooth Optimization

by

Alexander D. Joffe

Introduction

The purpose of this paper is to present statements of several new theorems with optimality conditions for nonsmooth optimization problems based on some recent developments in nonsmooth analysis. On the fundamental level, this is the theory of prederivatives [12] that enables to work with nonsmooth maps into infinite dimensional spaces and to attack higher order conditions problem. On the technical level, this is the notion of approximate subdifferential [13], [15] which provide for more selective necessary conditions in comparison with other derivative-like objects of nonsmooth analysis.

Since no details concerning approximate subdifferentials has been published thus far, we open the paper with a survey of their properties (§ 1). The most important among them are the following: (a) approximate subdifferentials are minimal (as sets) among other "generalized derivatives" satisfying some very natural requirements and (b) they admit a rich calculus (in certain respects even richer than the calculus of generalized gradients of Clarke).

In § 2 we consider the standard problem of mathematical programming with equality and inequality constraints and Lipschitz cost and constraint functions and maps. A statement of a Lagrange multiplier rule (L.M.R.) for the problem is presented. The novelty of the result is primarily that we no longer assume the range space of the equality constraint map finite dimensional. And it is stated in terms of approximate subdifferentials, hence being, in general, more selective than other results of such sort (thanks to the first property of approximate subdifferentials mentioned in the preceding paragraph).

The statement of the L.M.R. is followed by a discussion revealing an inherent and uncurable flaw of nonsmooth optimization problems. Figuratively speaking, the nonsmooth stationarity condition the L.M.R. incarnates is much less an "almost optimality" than its smooth counterpart. This creats a number of problems that seem to make nonsmooth optimization even more challenging theoretically though, probably, very difficult to approach from the computational viewpoint.

In § 3 we consider two optimal control problems, an abstract one and the standard problem with phase constraints, and for each of them we state a maximum principle which also seems to be most general among maximum principles for nonsmooth optimal control problems obtained by now.

In § 4 we return to the standard problem considered in § 2, this time to discuss second order conditions. Such a question seems to have never been touched upon in the literature (in connection with nonsmooth problems) though recently a different and very interesting idea was put forward by Aubin [1] (so far only for convex problems).

All spaces are assumed Banach and we use standard notation

for dual spaces, canonical pairing etc.. Necessary references and comments are gathered mostly at the ends of the sections. The paper is not a survey but rather a narrative about results obtained by the author. Therefore only those works are mentioned that played an actual role in the evolution of my understanding (which of course is a very subjective criterion).

§ 1. Approximate subdifferentials

1.1. Definition. Let f be a function on X finite at x. We set

$$d^-f(x;h) = \liminf_{\substack{t \searrow 0 \\ u \to h}} t^{-1}(f(x+tu) - f(x)),$$

$$\partial_\varepsilon^- f(x) = \{x^* \in X^* \mid \langle x^*, x \rangle \le d^-f(x;h) + \varepsilon \|h\|\}.$$

If $\varepsilon = 0$, we write $\partial^- f(x)$ (not $\partial_0^- f(x)$) and we set $\partial_\varepsilon^- f(x) = \emptyset$ if $|f(x)| = \infty$. The set $\partial_\varepsilon^- f(x)$ is called the <u>lower Dini</u> ε-<u>subdifferential</u> of f at x.

Let further $\mathcal{L}$ be a collection of subspaces of X. It is called <u>admissible</u> if (a) every $x \in X$ belongs to some $L \in \mathcal{L}$ and (b) for every two $L_1, L_2 \in \mathcal{L}$ there is $L \in \mathcal{L}$ containing both L_1 and L_2.

The collection $\mathcal{F}$ of all finite dimensional subspaces of X is an example of an admissible collection as well as the collection consisting of a single element X. Separable subspaces form another admissible collection etc..

Definition 1. Let $\mathcal{L}$ be an admissible collection of subspaces of X, let f be a function on X and

$$U(f,x,\varepsilon) = \{u \in X \mid \|u-x\| < \varepsilon, \ |f(u)-f(x)| < \varepsilon\}.$$

The set

$$\partial_A^{\mathcal{L}} f(x) = \bigcap_{\substack{L \in \mathcal{L} \\ \varepsilon > 0}} \overline{\bigcup_{u \in U(f,x,\varepsilon)} \partial_\varepsilon^- f(u)}$$

is called the (broad analytic) <u>approximate</u> $\mathcal{L}$-<u>subdifferential</u> of f at x. (The bar denotes the weak* closura) Let us agree to denote approximate $\mathcal{F}$-subdifferentials by $\partial_A f(x)$, without any superscript.

The words in parentheses refer to [15] where a number of other approximate subdifferentials were introduced.

An important property of approximate subdifferentials is that within certain limits they do not actually depend on $\mathcal{L}$. Let us say that X is a <u>weakly trustworthy space</u> (WT-space) if for any two l.s.c. functions f_1 and f_2 on X, any $\varepsilon > 0$ and any weak* neighbourhood V about the origin in X^*

$$\partial_\varepsilon^-(f_1 + f_2)(x) \subset \bigcup_{x_i \in U(f_i,x,\varepsilon)} (\partial_\varepsilon^- f_1(x_1) + \partial_\varepsilon^- f_2(x_2) + V).$$

Proposition 1. <u>Any finite dimensional</u>, <u>any separable Banach space and</u>, <u>more generally</u>, <u>any Banach space with an equivalent Gâteaux differentiable norm is a WT-space</u>.

Loosely speaking, a WT-space is such a space that admits a good calculus of Dini ε-subdifferentials of functions thereon. The following theorem gives a formal characterization of the above mentioned property of approximate subdifferentials.

Theorem 1. <u>Let</u> $\mathcal{L}$ <u>be an admissible family of WT-subspaces of</u> X. <u>Then</u>

$$\partial^{z}_{A} f(x) = \partial_A f(x)$$

for any l.s.c. function f *and any* x.

1.2. Calculus. In this section we shall list some important analytic properties of approximate subdifferentials.

Theorem 2. *The set-valued map* $x \to \partial_A f(x)$ *is u.s.c. in the sense that*

$$\partial_A f(x) = \bigcap_{\varepsilon > 0} \overline{\bigcup_{u \in U(f,x,\varepsilon)} \partial_A f(u)}.$$

In other words, if $x_n \to x$, $f(x_n) \to f(x)$ and x_n^* belongs to $\partial_A f(x_n)$, then any weak* limit point of the sequence $\{x_n^*\}$ belongs to $\partial_A f(x)$.

Theorem 3. *If* f *is strictly differentiable at* x, *then* $\partial_A f(x) = \{f'(x)\}$;

if f *is convex and continuous at some point, then* $\partial_A f(x) = \partial f(x)$, *the subdifferential of* f *at* x *in the sense of convex analysis*;

if f *is Lipschitz near* x, *then the convex closure of* $\partial_A f(x)$ *coincides with* $\partial_c f(x)$, *the Clarke generalized gradient of* f *at* x.

An approximate subdifferential may be noticeably smaller than the corresponding Clarke generalized gradient, in particular it is typically a nonconvex set. For instance, if $X = R^n$ and f is a concave continuous function, then $\partial_A f(x)$ is the collection of all limits $\lim f'(x_k)$, where $x_k \to x$ and f is differentiable at x_k.

An interesting question arising in connection with the last

statement of Theorem 3 is: is the equality $\partial_A f(x) = \partial_c f(x)$ (f is Lipschitz) an exotic property or not? In all verifiable examples we know this equality holds on a massive set. So it may well happen that, inspite of Theorem 3, the equality above is a generic property.

But whatever answer this question has, the property of being smaller is an important advantage from the optimizational viewpoint. Indeed, both inclusions $0 \in \partial_c f(x)$ and $0 \in \partial_A f(x)$ are necessary for f to have a local minimum at x but the second possess a greater selective power. For example, if $f(x) = -\|x\|$, then $0 \in \partial_c f(0)$ but $0 \notin \partial_A f(0)$.

We have to note also that it can be likewise shown that, in the finite dimensional space, any approximate subdifferential is not greater than the corresponding derivative container of Warga.

Theorem 4. Assume that the functions f and g are l.s.c. and one of them is Lipschitz near z. Then

$$\partial_A (f + g)(z) \subset \partial_A f(z) + \partial_A g(z).$$

If both f and g are Lipschitz near z, then

$$\partial_A (f \vee g)(z) \subset \bigcup_{0 \leq \alpha \leq 1} \partial_A (\alpha f + (1-\alpha) g)(z).$$

This theorem must look very surprising for anyone acquainted with convex and nonsmooth analysis. Indeed, so far any such result was essentially based on convexity and this is the first time that such inclusions are proved to be valid for nonconvex objects. The proof based on what could be called a version of the penalty function method is rather very technical though it incorporates some very simple observations.

One of them is that in case $\dim X < \infty$ the inclusion $0 \in \partial_{\varepsilon}^{-} f(z)$ implies that $f(x) + \delta \| x-z \|$ attains a local minimum at z for any $\delta > \varepsilon$. Another and very interesting consequence of this observation is that the approximate subdifferential is in a sense the smallest possible "generalized derivative". The proposition below is the exact statement of the fact for the finite dimensional case. Things are more complicated if $\dim X = \infty$ but definitely the same is true for Lipschitz functions on arbitrary spaces.

Proposition 2. <u>Assume that a procedure is given that associates with any function</u> f <u>on a finite dimensional space</u> X <u>and with any</u> $x \in X$ <u>a set</u> $\partial f(x) \subset X^*$ <u>in such a way that</u>

(a) <u>the set-valued map</u> $x \to \partial f(x)$ <u>is u.s.c. in the sense of Theorem</u> 2;

(b) <u>if</u> f <u>is convex continuous, then</u> $\partial f(x)$ <u>is the subdifferential of</u> f <u>in the sense of convex analysis</u>;

(c) <u>if</u> f <u>attains a local minimum at</u> x, <u>then</u> $0 \in \partial f(x)$;

(d) $\partial (f+g)(x) \subset \partial f(x) + \partial g(x)$ <u>whenever</u> g <u>is convex continuous</u>.

<u>Then</u> $\partial_A f(x) \subset \partial f(x)$ <u>for any</u> f <u>and</u> x.

The proof of the fact is very simple, so we are able to present it completely.

If $x^* \in \partial_A f(x)$, then $x^* = \lim x_k^*$, where $x_k^* \in \partial_{\varepsilon_k}^{-} f(x_k)$ $\varepsilon_k \to 0$, $x_k \to x$, $f(x_k) \to f(x)$. It follows that

$$g_k(x) = f(x) - \langle x_k^*, x-x_k \rangle + 2\varepsilon_k \| x-x_k \|$$

attains a local minimum at x_k. According to (c), $0 \in \partial g_k(x_k)$

and according to (b) and (d), $0 \in \partial f(x_k) - x_k^* + 2\varepsilon_k B$ (B being the unit ball in X^*) which is the same as $x_k^* \in f(x_k) + 2\varepsilon_k B$. It remains to apply (a) to conclude that $x^* \in \partial f(x)$.

Any reasonable "differentiation" should have properties (b) through (d), so only (a) is actually an assumption. On the other hand, the lower semicontinuity assumption (a) is a very weak nonsmooth replacement for the strict (or strong) differentiability property which is so important in many situations.

We conclude this section with a chain rule for approximate subdifferentials. But first we recall (see [12]) that

a set-valued map $\mathcal{A}$ from X into another Banach space Y is <u>homogeneous</u> if $0 \in \mathcal{A}(0)$ and $\mathcal{A}(\lambda x) = \lambda \mathcal{A}(x)$ for all $\lambda > 0$, $x \in X$; $\mathcal{A}$ is <u>bounded</u> if there is $k \geqslant 0$ such that $\|y\| \leq k\|x\|$ whenever $y \in \mathcal{A}(x)$;

a homogeneous set-valued map $\mathcal{A}$ from X into Y is a <u>strict prederivative</u> of a map $F: X \to Y$ at z if

$$F(x+h) \subset F(x) + \mathcal{A}(h) + r(x,h)\|h\| B_Y ,$$

where $r(x,h) \to 0$ if $x \to z$, $h \to 0$ and B_Y is the unit ball in Y.

It must be noted that a map having a bounded strict prederivative at a point is necessarily Lipschitz near the point.

Theorem 5. <u>Let</u> $G: X \to Y$ <u>have a bounded strict prederivative at</u> z <u>with norm compact values, and let</u> $g(y)$ <u>be a function on</u> Y <u>which is Lipschitz near</u> $G(z)$. <u>We set</u> $f(x) = g(G(x)) = (g \circ G)(x)$. <u>Then</u>

$$\partial_A f(z) \subset \bigcup_{y^* \in \partial_A g(G(z))} \partial_A (y^* \circ G)(z).$$

It is worth noting that this chain rule is sharper that the one we have for generalized gradients of Clarke that necessarily involves the convex closure operation. Using Theorem 5, we can obtain a more precise formula for the Clarke generalized gradient of a composition:

$$\partial_c f(z) \subset \overline{\text{conv}} \bigcup_{y^* \in \partial_A g(G(z))} \partial_A (y^* \circ G)(z)$$

(compare this to the known formula [5], [12] obtained from the above one by replacing ∂_A by ∂_c).

1.3. Comments. Approximate subdifferentials were introduced in [13] as an infinite dimensional generalization of "lower generalized derivatives" introduced by Mordukhovich [21]. The original definition of the latter was essentially finite dimensional and heavily geometrical. An equivalent analytic definition (using lower Dini derivatives) was found much later. But first attempts to extend it to infinite dimensional case were not fully successful. Constructions offered in [12] (m-subdifferentials) and [20] could work only in spaces with equivalent Gâteaux or Fréchet differentiable norm respectively.

The definition of approximate subdifferentials given here works in all Banach spaces, in fact even in locally convex spaces [13]. In spaces with equivalent Gâteaux or Fréchet differentiable norm approximate subdifferentials coincide with the objects introduced in [12] and [20]. The proof of the fact based on Proposition 1 is not trivial. In general, the possibility to choose an admissible family without changing the result given by Proposition 1 proved to be very convenient and usable.

Recently Kruger showed [18] that the first part of Theorem 4 is valid under weaker assumptions. It was done for the objects introduced in [20] that make sense only in spaces with an equivalent Fréchet differentiable norm but the extension to general approximate subdifferentials presents no difficulty.

We refer to [15] for proofs and more details concerning approximate subdifferentials.

§ 2. Lagrange multiplier rule

2.1. Statement of the problem and the theorem. The problem we shall consider in this section is the standard problem of mathematical programming:

$$\text{minimize} \quad f_o(x) \tag{1}$$

subject to

$$f_i(x) \leq 0, \quad i = 1,\dots,n \; ; \quad F(x) = 0. \tag{2}$$

The functions f_i and the map $F: X \to Y$ will be assumed Lipschitz near a supposed solution z (though the theorem below is valid if one of the functions is only l.s.c.).

Even in the smooth case we need to impose some restrictions on F if the range space of F is not finite dimensional. Usually it is assumed that the range of $F'(z)$ (the derivative of F at z) is closed and/or has finite codimension. We start by introducing a nonsmooth analogue for the latter property.

Let $\mathcal{A}$ be a bounded homogeneous set-valued map from X into Y. We set

$$s(y^*,x) = \sup\{\langle y^*,y\rangle \mid y \in \mathcal{A}(x)\},$$

$$C(\mathcal{A},V^*) = - \sup_{\substack{\|y^*\|=1 \\ y^* \in V^*}} \; \inf_{\|x\|\leq 1} s(y^*,x)$$

(see [12]). It is obvious that $C(\mathcal{A},V^*) \geqslant 0$ (since $s(y^*,0) = 0$).

We shall say that $\mathcal{A}$ has the finite codimension property if there is a weak* closed subspace $V^* \subset Y^*$ of finite codimension such that $C(\mathcal{A},V^*) > 0$.

A linear operator (regarded as a set-valued map) has the finite codimension property if and only if the codimension of the range of the operator is finite.

Theorem 6. <u>Assume that</u>

(a) <u>there is an equivalent Gâteaux differentiable norm in</u> Y;

(b) F <u>has a norm-compact-valued bounded strict prederivative at</u> z <u>having the finite codimension property</u>.

<u>If, under these assumptions,</u> z <u>is a local solution to</u> (1), (2), <u>then there are</u> $\lambda_0 \geqslant 0,\ldots,\lambda_n \geqslant 0$, $y^* \in Y^*$ <u>not all equal to zero and such that</u>

$$0 \in \partial_A(\lambda_0 f_0 + \ldots + \lambda_n f_n + y^* \circ F)(z) \tag{3}$$

$$\lambda_i f_i(z) = 0 \ , \quad i = 1,\ldots,n.$$

2.2. A sketch of the proof. Recall [11] that F is <u>regular</u> at z if there is $k > 0$ such that

$$\rho(x, \mathcal{L}(F,z)) \leq k \| F(x) - F(z) \|$$

for all x sufficiently close to z. Here

$$\mathcal{L}(F,z) = \{ x \mid F(x) = F(z) \}$$

and ρ stands for the distance. It is also said that F is <u>locally surjective</u> at z if the image under F of any neighbourhood of z contains a neighbourhood of F(z). Both properties are closely connected ([6], [12]).

The proof of the theorem is based on the following

Proposition 3. <u>Let the assumptions</u> (a), (b) <u>of the theorem be valid</u>. <u>We set</u>

$$c^* = \inf\{\|x^*\| \mid x^* \in \partial_A(y^* \circ F)(z),\ \|y^*\| = 1\}.$$

<u>If</u> $c^* > 0$, <u>then</u> F <u>is regular and locally surjective at</u> z. <u>If</u> $c^* = 0$, <u>then there is</u> $y^* \in Y^*$, $\|y^*\| = 1$ <u>such that</u>

$$0 \in \partial_A(y^* \circ F)(z).$$

This proposition is another version of the "controllability-multiplier rule alternative" well known in smooth and less general nonsmooth situations (see [25] for example). Once it has been established, the proof of the theorem runs as follows.

If $c^* = 0$, we set $\lambda_o = \ldots = \lambda_n = 0$ and take y^* from Proposition 3. If $c^* > 0$ then by the reduction theorem of [11] there is $k > 0$ such that z is a local minimum of the function

$$g(x) = \max_{0 \le i \le n} f_i(x) + k\|F(x)\|$$

(we have set $f_o(z) = 0$ for simplicity). Therefore $0 \in \partial_A g(z)$ and it remains to apply Theorem 5 to complete the proof.

2.3. A remark about nonsmooth multiplier rules. It is customary and natural to test the "quality" of one or another nonsmoothness theorem by applying it in a smooth situation: if the corresponding result is implied without very restrictive additional assumptions, the theorem is considered rather good.

In this sense Theorem 6 seems to be quite good: it does imply the "smooth" multiplier rule and the additional assumption that an equivalent Gâteaux differentiable norm exists in Y

is not that much restrictive. Indeed, any separable space has this property.

But we wish to look at the problem from another point of view. For simplicity, let us consider only problems with equality constraints. As well known, if F is regular at z, then the smooth Lagrange multiplier rule

$$0 = f'(z) + (y^* \circ F')(z)$$

is equivalent to the fact that, for any $\varepsilon > 0$, z is a local solution to the problem

$$\text{minimize} \quad f(x) + \varepsilon \|x-z\|, \quad \text{s.t.} \quad F(x) = 0. \tag{4}$$

What property of nonsmooth problems the nonsmooth multiplier rule is equivalent to ? We cannot answer this question but (3), now assuming the form

$$0 \in \partial_A (f + (y^* F))(z) \tag{5}$$

definitely does not imply that z is a solution to (4).

There are some other nice properties of smooth L.M.R. that its nonsmooth analogue does not have. It is not difficult, for example, to construct a regular problem with only one local and global extremum and such that the L.M.R. holds at every admissible point. In the smooth situation this is impossible. It can be also shown [6] that the nonsmooth L.M.R. is not invariant with respect to nonsingular Lipschitz transformations.

2.4. Comments. Theorem 6 (and Proposition 3) with Clarke generalized gradients instead of approximate subdifferentials was established in [14]. (The proof for approximate subdifferentials is the same in principle though it differs in certain details.) Till then only nonsmooth problems with finitely many equality constraints were considered. Moreover, the first nonsmooth L.M.R. were obtained in a weaker form ([3], [9], [25]):

$$0 \in \lambda_o \partial f_o(z) + \ldots + \lambda_n \partial f_n(z) + \sum \mu_i \partial (e_i \circ F)(z), \qquad (6)$$

(where e_i is a basis in Y and ∂ stands for the generalized derivative used).

In the form (3), the nonsmooth L.M.R. was first established in [10] for completely finite dimensional situation ($\dim X < \infty$) and later in [6] in a somewhat stronger form and for arbitrary X. A multiplier rule involving approximate subdifferentials was also obtained in [19] but in the form (6), with finite dimensional X and Y having an equivalent Fréchet differentiable norm.

In some works ([3],[6], [14]) a nonfunctional constraint $x \in S$ is also included into the problem. Using approximate subdifferentials does not creat additional difficulties in handling such a constraint. We have not included it only to avoid introducing and discussing a number of notions of tangency and normalcy connected with approximate subdifferentials.

As to Proposition 3, it is completely along the lines of [12] and can be stated in terms of "approximate coderivatives" and dual Banach constants.

§ 3. Maximum principle in nonsmooth optimal control problems

3.1. An abstract problem. We proceed to consider a more general problem embracing both the problem considered in the preceding section and the optimal control problem. It is stated as follows

$$\text{minimize} \quad f_o(x) \tag{7}$$

subject to

$$f_i(x) \leq 0,\ i = 1,\dots,n; \quad F(x,u) = 0, \quad u \in U \ . \tag{8}$$

As before, X and Y are Banach spaces, the functions f_i and the map F are Lipschitz (the latter as a function of x) near $z \in X$ satisfying (7) and U is an arbitrary set. We shall say that (z,v) is a local solution of (7), (8) if it is admissible (i.e. satisfies (8)) and no other admissible pair (x,u) with x sufficiently close to z gives a smaller value to the cost function f_o.

Fix a finite collection $\{u_1,\dots,u_m\}$ of elements of U and set

$$\Phi(a,x) = F(x,v) + \sum_{j=1}^{m} \alpha_j (F(x,u_j) - F(x,v)),$$

where $a = (\alpha_1,\dots,\alpha_m)$. By Σ^m we denote the m-simplex $\{a \in R^m \mid \alpha_j \geq 0,\ \Sigma\alpha_j = 1\}$. We impose the following requirements on F:

(H_1) the map $F(.,v)$ has a norm-compact-valued bounded strict prederivative satisfying the finite codimension property;

(H_2) for any finite collection $\{u_1,\dots,u_m\}$ of elements of U and for any $\delta > 0$ there is a map $v_\delta(a,x)$ from $(\delta\Sigma^m) \times B(z,\delta)$ (the δ-ball around z) into U such that $v_\delta(0,x) = v$ and

$$\| \Phi(a,x) - \Phi(a',x') - F(x,v_\delta(a,x)) + F(x',v_\delta(a',x')) \|$$

$\leqslant q(\delta)(\|x-x'\| + \|a-a'\|), \ \forall\ a,a' \in \delta\Sigma^m, \quad x,x' \in B(z,\delta),$

where $q(\delta) \to 0$ as $\delta \to 0$.

The second hypothesis in the present form appeared first in [17] but its roots lie in earlier works of Halkin and Neustadt [8], [22]. It is aimed at including such phenomena as sliding regimes into variational analysis without explicit introducing the relaxed problem. The hypothesis is satisfied if F is convex in u in the sense that $F(x,U)$ is a convex set for any x sufficiently close to z.

Theorem 7. <u>Assume hypotheses</u> (H_1), (H_2) <u>and assume that</u> Y <u>has an equivalent Gâteaux differentiable norm</u>. <u>If</u> (z,v) <u>is a local solution of</u> (7), (8), <u>then there are</u> $\lambda_0 \geqslant 0, \ldots, \lambda_n \geqslant 0$, $y^* \in Y^*$ <u>not all equal to zero and such that</u>

$$0 \in \partial_A(\lambda_0 f_0 + \ldots + \lambda_n f_n + (y^* \circ F)(\cdot,v))(z),$$

$$(y^* \circ F)(z,v) = \max_{u \in U} (y^* \circ F)(z,u),$$

$$\lambda_i f_i(z) = 0, \quad i = 1,\ldots,n.$$

It is clear that Theorem 6 is a particular case of Theorem 7 (if U is a singleton). But the proof of Theorem 7 follows from Theorem 6 augmented by Dubovitzkii-Miljutin's method of projective systems of necessary conditions [7].

Therefore this theorem lies between Lagrange multiplier rule and the maximum principle (cf. [17, Ch. 5]). The role of the theorem is that it enables to prove maximum principles for different optimal control problems (say, including partial differential or integral equations) that admit relaxation procedures.

3.2. Optimal control with state constraints. The problem we

shall consider next is the standard optimal control problem with state constraints:

$$\text{minimize} \quad f(x(0),x(1)) \tag{9}$$

subject to

$$\dot{x} = \varphi(t,x,u), \quad u \in U(t); \tag{10}$$

$$g_i(t,x) \leq 0, \quad t \in [0,1], \quad i = 1,\ldots,k; \tag{11}$$

$$h(x(0),x(1)) = 0 \tag{12}$$

(of course without differentiability assumptions).

We shall use the standard terminology and notation accepted in the theory of optimal control without further explanations. Note only that x is an n-dimensional vector, $U(t)$ is an arbitrary set in R^r and $h: R^n \times R^n \to R^m$.

The hypotheses we shall adopt are the following:

(H_3) f and h are Lipschitz in a neighbourhood of $(z(0),z(1))$ (in fact, f may be l.s.c.);

(H_4) for any measurable $u(t)$ the map $t \to \varphi(t,x,u(t))$ is measurable;

(H_5) for any admissible control $u(t)$ (i.e. such that $u(t)$ $U(t)$ and measurable) there are $\varepsilon(t) > 0$ and $k(t) \geq 0$ (both measurable) such that

$$\|\varphi(t,x,u(t)) - \varphi(t,x',u(t))\| \leq k(t)\,\|x-x'\|$$

whenever $\|x-z(t)\| < \varepsilon(t)$, $\|x'-z(t)\| < \varepsilon(t)$, $t \in [0,1]$; for $u(.) = v(.)$ these functions can be chosen in such a way that $k(t)$ be summable and $\varepsilon(t) = \varepsilon > 0$ a.e.;

(H_6) the functions $g_i(t,x)$ are jointly continuous and Lipschitz in x in a neighbourhood of the graph of $z(.)$ $(t \in [0,1])$.

The hypotheses of course are very loose.

As usual we set

$$H(t,p,x,u) = \langle p, \varphi(t,x,u) \rangle \qquad (p \in R^n),$$

$$T_i = \{ t \in [0,1] \mid g_i(t,z(t)) = 0 \}.$$

We set also

$$\bar{\partial}_A g_i(t,x) = \bigcap_{\varepsilon > 0} \overline{\bigcup_{|\tau - t| < \varepsilon} \partial_A g(\tau,x)} ,$$

where ∂_A here and below refers to the dependence on x.

Theorem 8. <u>Assume</u> (H_3) - (H_6). <u>If</u> $(z(.),v(.))$ <u>is a local solution of</u> (9) - (12), <u>then there are</u> $\lambda \geq 0$, <u>a map</u> $p(.): [0,1] \to R^n$ <u>of bounded variation, vectors</u> $l \in R^m$ <u>and</u> $q \in R^n$ <u>and nonnegative Radon measures</u> μ_i <u>supported on</u> T_i $(i = 1,\dots,k)$ <u>such that</u>

$$\lambda + \|p(t)\| + \sum_i \|\mu_i\| + \|l\| > 0, \qquad \forall t,$$

$$p(t) \in q + \int_t^1 [\partial_A H(\tau,p(\tau),z(\tau),v(\tau))d\tau - \sum_{i=1}^k \bar{\partial}_A g_i(\tau,z(\tau))d\mu_i(\tau)] ,$$

$$(p(0),-q) \in \lambda\,\partial_A f(z(0),z(1)) + \partial_A(l \circ h)(z(0),z(1))$$

<u>and</u>

$$H(t,p(t),z(t),v(t)) \geq H(t,p(t),z(t),u(t))$$

<u>for any admissible</u> $u(.)$.

<u>If in addition, the graph of the set-valued map</u> $U(t)$ <u>is Souslin and the map</u> $(t,u) \to \varphi(t,z(t),u)$ <u>is</u> $\mathcal{L} \otimes \mathcal{B}$<u>-measurable then the last inequality can be replaced by</u>

$$H(t,p(t),z(t),v(t)) = \max_{u \in U(t)} H(t,p(t),z(t),u) \quad \underline{a.e.}.$$

The proof of the theorem includes reformulation of the problem in terms of the abstract problem (7), (8), verification of the hypotheses (H_1) and (H_2) and backward reformulation of the abstract maximum principle given by Theorem 7.

3.3. Comments. First maximum principles for nonsmooth optimal control problems under general assumptions were obtained by Clarke [2], [4] and Warga [23], [24]. Clarke considered problems without state constraints but under very weak hypotheses about the components of the problem. It was Clarke who observed that no continuity in t and u is needed in the right side of the differential equation.

Theorem 8 generalizes the results of both. (It is actually stronger for we consider approximate subdifferentials.) The assumptions we impose are even weaker than the "minimal hypotheses" of Clarke [4].

Proofs of Theorems 7 and 8 will appear in [14] (again with Clarke generalized gradients and again it is not a difficult matter to adjust the proof for the approximate subdifferentials). The method used to prove the theorems is very close to that developed in [17, Ch. 5]. In fact, some technical simplifications have allowed to make the proofs even shorter.

We note fimally that Mordukhovich's paper in which his generalized derivatives were originally introduced was also devoted to optimal control problems but only the cost function and the endpoint constraint map were assumed nonsmooth there.

§ 4. Second order conditions

In this section we return to the problem (1), (2) and state a theorem containing a necessary and a sufficient second order conditions for z to be a local solution. But this will be done under different assumptions:

(H_7) the functions $f_o,\dots,f_n$ and the map F are Fréchet differentiable and their derivatives are Lipschitz continuous near z;

(H_8) $F'(z)$ maps X onto Y.

According to the first hypothesis, the problem may be qualified as "second order nonsmooth". Motivations for the hypothesis are quite obvious. Since necessary conditions obtained under "first order" nonsmoothness assumptions (as in § 2) may be so far from real optimality, subtle second order tests will be just useless.

The second hypothesis is rather usual for second order necessary conditions even in the smooth case [11] .

Let

$$\mathcal{L}(\lambda_o,\dots,\lambda_n,y^*;x) = \lambda_o f_o(x) + \dots + \lambda_n f_n(x) + (y^* \circ F)(x)$$

be the Lagrangian of the problem, and for the sake of brevity, let us denote collections of multipliers $(\lambda_o,\dots,\lambda_n,y^*)$ by q, so that the Lagrangian may be written as $\mathcal{L}(q,x)$.

We set (see [12])

$$s_\varepsilon(q,e,h) = \sup_{\substack{\|x-z\|<\varepsilon \\ \|x+te-z\|<\varepsilon \\ 0<t<\varepsilon}} t^{-1}(\mathcal{L}'(q,x+te)h - \mathcal{L}'(q,x)h)$$

$$s(q,e,h) = \lim_{\varepsilon\to 0} s_\varepsilon(q,e,h)$$

$$= \limsup_{\substack{x \to z \\ t \searrow 0}} \; t^{-1}(\mathcal{L}'(q,x+te)h - \mathcal{L}'(q,x)h)$$

(Here and below the prime denotes the derivative in x). In other words, $s(q,e,h)$ (as a function of e) is Clarke's directional derivative of $x \to \mathcal{L}'(x)h$ at z.

If z is a local solution of (1), (2), then the set

$$Q = \{ q \mid \mathcal{L}'(q,z) = 0, \quad \lambda_i \geqslant 0, \quad \lambda_o + \ldots + \lambda_n = 1$$

of normalized Lagrange multipliers (there is no loss of generality in assuming that $f_i(z) = 0$, $i = 1,\ldots,n$) is nonempty thanks to (H_8). We set also

$$Q_\varepsilon = \{ q \mid \| \mathcal{L}'(q,z) \| \leqslant \varepsilon \;, \; \lambda_i \geqslant 0 \;, \; \lambda_o + \ldots + \lambda_n = 1 \} .$$

Let finally,

$$K = \{ h \in X \mid f_i'(z)h \leqslant 0, \quad i = 0,\ldots,n; \quad F'(z)h = 0 \}$$

be the crictical cone at z.

Theorem 9. <u>Assume</u> (H_7), (H_8). <u>If</u> z <u>is a local solution of</u> (1), (2), <u>then</u>

$$\sup_{q \in Q_\varepsilon} s_\varepsilon(q,h,h) \geqslant 0, \qquad \forall\, h \in K, \qquad \forall \varepsilon > 0.$$

<u>On the other hand, if there is</u> $k > 0$ <u>such that</u>

$$\inf_{q \in Q} s(q,-h,h) \leqslant -k \|h\|, \qquad \forall\, h \in K,$$

<u>then</u> z <u>is an isolated local solution of</u> (1), (2).

<u>If in addition, the map</u> $x \to F'(x)$ <u>has a bounded strict prederivative at</u> z <u>with compact values (in the strong operator topology), then the necessary condition above is valid</u>

<u>also for</u> $\varepsilon = 0$. <u>This is true in particular if</u> $\dim Y < \infty$.

A version of this theorem for a more general problem (semi-infinite programming) but in the finite dimensional situation will be proved in [16]. As to the proof of Theorem 9, it will be published elsewhere. It is worth saying, however, that the proof is very similar to what was done in the smooth case in [11].

The main innovation of the theorem is the appearance of the "trisublinear function" $s(q,e,h)$ (cf. [12]) instead of the Hessian of the Lagrangian in the smooth case. Notice that in this case $s(q,h,h) = - s(q,-h,h)$ is just the Hessian, i.e. the theorem withstands the smoothness test !

References

1. J.-P. Aubin, Lipschitz behavior of solutions to convex minimization problems, Working Paper A-2361, IIASA, Laxanburg 1981.

2. F.H. Clarke, A general control problem, in <u>Calculus of Variations and Optimal Control</u>, D.L. Russel, editor, Academic Press 1976, pp. 257 - 278.

3. _________, A new approach to Lagrange multipliers, Math. Operation Res., 1 (1976), 165-174.

4. _________, The maximum principle under minimal hypotheses, SIAM J. Control Optimization 14 (1976), 1078 - 1091.

5. _________, Generalized gradients of Lipschitz functionals, Adv. Math.

6. A.V. Dmitruk, A.A. Miljutin and N.P. Osmolovskii, The Ljusternik theorem and the theory of extremum, Uspehi Mat. Nauk 35:6 (1980), 11 - 46.

7. A.Ya. Dubovitskii and A.A. Miljutin, Translation of Euler equations, J. Computational Math. and Mathematical Physics, 9 (1969), 1263 - 1284.

8. H. Halkin, Optimal Control as programming in infinite dimensional spaces, in Calculus of Variations, Classical and Modern, Edizioni Cremonese, Roma, 1966, 179 - 192.

9. ______ , Mathematical programming without differentiability, in Calculus of Variations and Optimal Control, D.L. Russell, ed., Academic Press, 1976, 279 - 288.

10. J.-B. Hiriart-Urruty, Refinements of necessary optimality conditions in nondifferentiable programming, Appl. Math. Optim. 5 (1979), 63 - 82.

11. A.D. Ioffe, Necessary and sufficient conditions for a local minimum, SIAM J. Control Optimization 17 (1979), 245-288.

12. _____ , Nonsmooth analysis: differential calculus of non-differentiable mappings, Trans. Amer. Math. Soc. 266 (1981),1-56.

13. _____ , Sous-différentielles approchées de fonctions numériques, C.R. Acad. Sci. Paris (1981),

14, _____ , Necessary conditions in nonsmooth optimization, Math. Operation Res., to appear.

15. _____ , Approximate subdifferentials of nonconvex functions, Cahiers de Math. de la Decision, CEREMADE, Paris 1981

16. _____ , Second order conditions in nonlinear nonsmooth semi-infinite programming, Intern. Symp. on Semi-Infinite Programming, Austi, Texas, September 1981.

17. A.D. Ioffe and V.M. Tikhomirov, Theory of Extremal Problems, Nauka, Moscow, 1974, North Holland, 1979.

18. A. Kruger, Calculus of generalized differentials, to appear

19. _____ , Generalized differentials of nonsmooth functions and

necessary conditions for an extremum, to appear.

20. A. Kruger and B. Mordukhovich, Extremal points and the Euler equation in nonsmooth optimization problems, Dokl. Acad. Nauk BSSR 24 (1980), 684-687.

21. B. Mordukhovich, Maximum principle in the optimal time control problem with nonsmooth constraints, Appl. Math. Mech. 40 (1976), 1014-1023.

22. L.W. Neustadt, Optimization, Princeton Univ. Press 1976.

23. J. Warga, Necessary conditions without differentiability assumptions in optimal control, J. Diff. Eqs. 18 (1975), 41 - 62.

24. _____ , Derivative containers, inverse functions and controllability, in Calculus of Variations and Control Theory, D.L. Russell, editor, Academic Press, 1976.

25. _____ , Controllability and a multiplier rule for nondifferentiable optimization problems, SIAM J. Control Optimization 16 (1978), 803 - 812.

CONTROLE OPTIMAL DE SYSTEMES A ETATS MULTIPLES

J.L. LIONS

Collège de France et INRIA

1. Introduction

1.1. Position du problème

Soit Ω un ouvert borné de $\mathbb{R}^n$, n = 2 ou 3, de frontière Γ assez régulière.

On considère dans Ω l'équation

(1.1) $$- \Delta z - z^3 = v$$

avec la condition aux limites

(1.2) $$z = 0 \text{ sur } \Gamma.$$

La fonction v est le *contrôle* et "la" fonction z est l'*état*. Mais l'équation (1.1) avec (1.2) *admet en général une infinité de solutions*. De manière précise, on sait (Bahri [1]) que pour un ensemble *dense dans* $L^2(\Omega)$ de fonctions v, le problème (1.1) (1.2) *admet une infinité de solutions*, et on *conjecture* (Bahri - Berestycki [1]) qu'il en est toujours ainsi, i.e. quel que soit v dans $L^2(\Omega)$. On dit, en conséquence, que l'on a affaire à un *système à états multiples*. □

Pour étudier le contrôle d'un tel système, plutôt que d'essayer de suivre des "branches" de telle solution particulière de (1.1) (1.2), on va *changer de point de vue*. □

On considère *a priori* l'ensemble des couples $\{v,z\}$ tels que

(1.3) $$v \in L^2(\Omega) \quad , \quad z \in L^6(\Omega)$$

et tels que (1.1) (1.2) aient lieu.

La condition (1.2) est prise au sens suivant : il résulte de (1.1) que

$$- \Delta z = v + z^3 \quad , \quad v + z^3 \in L^2(\Omega)$$

et donc, si l'on considère $v + z^3$ comme donné dans $L^2(\Omega)$, on obtient :

(1.4) $z \in H^2(\Omega) \cap H_0^1(\Omega)$ [1] .

On considère alors la *fonction coût*

(1.5) $$J(v,z) = \frac{1}{6} ||z-z_d||^6_{L^6(\Omega)} + \frac{N}{2} ||v||^2_{L^2(\Omega)} .$$ □

Remarque 1.1.

En théorie habituelle du contrôle optimal des systèmes distribués, (cf. J.L. Lions [1]), l'état est donné par la solution de l'équation d'état, par exemple, si l'on considère l'équation

(1.6) $$\begin{cases} - \Delta y + y^3 = v \\ y|_\Gamma = 0 \end{cases}$$

elle *admet une solution unique* dans $H_0^1(\Omega)$ si $v \in L^2(\Omega)$; soit $y(v)$ cette solution; comme $n \leq 3$, on a (Sobolev)

(1.7) $$H_0^1(\Omega) \subset L^6(\Omega)$$

et l'on considère la fonction coût (par exemple)

(1.8) $$J(v) = \frac{1}{6} ||y(v)-z_d||^6_{L^6(\Omega)} + \frac{N}{2} ||v||^2_{L^2(\Omega)} .$$

La situation dans (1.5) est *différente*, puisqu'il est très difficile, sinon impossible, de considérer z comme fonction de v. □

Remarque 1.2.

La fonction coût (1.5) est prise comme "modèle", ainsi que l'équation (1.1). Tout cela peut être considérablement généralisé. De même peut-on remplacer la condition de Dirichlet (1.2) par d'autres conditions aux limites. □

---

(1) Avec les notations habituelles des espaces de Sobolev; $H^m(\Omega)$ est l'espace des fonctions $L^2(\Omega)$ telles que toutes leurs dérivées d'ordre $\leq m$ soient dans $L^2(\Omega)$ et $H_0^m(\Omega)$ l'adhérence de $\mathcal{D}(\Omega) = C_0^\infty(\Omega)$ dans $H^m(\Omega)$.

Remarque 1.3.

On peut également, dans l'esprit de cette note, considérer les couples $\{v,z\}$ tels que l'on ait (1.1) *sans condition aux limites*, ou avec des conditions aux limites "insuffisantes" ($z = 0$ sur une *partie* Γ_0 de Γ), ou, au contraire, avec des conditions aux limites "surabondantes" telles que, par exemple, $z = \frac{\partial z}{\partial \nu} = 0$ sur Γ (problème de Cauchy).

Des situations de ce genre ont été étudiées dans J.L. Lions [2] et feront l'objet d'une présentation systématique dans J.L. Lions [5]. □

Les contraintes.

On introduit maintenant :

(1.9) $\mathcal{U}_{ad}$ = ensemble convexe fermé non vide de $L^2(\Omega)$

et on va *supposer* (cf. Remarque 1.4 ci-après)

(1.10) il existe un couple $\{v_0,z_0\}$ vérifiant (1.1) (1.2) (1.3) et tel que $v_0 \in \mathcal{U}_{ad}$

On va désigner par $E(\mathcal{U}_{ad})$ l'ensemble

(1.11) $E(\mathcal{U}_{ad}) = \{\{v,z\}|,\ v,z$ satisfont à (1.1) (1.2) (1.3) et $v \in \mathcal{U}_{ad}\}$.

On considère alors *le problème de contrôle optimal* :

(1.12) minimiser $J(v,z)$ sur $E(\mathcal{U}_{ad})$.

Remarque 1.4.

Si la conjecture de Bahri-Berestycki est vérifiée, alors l'hypothèse (1.10) est *toujours* vérifiée. □

1.2. Motivation

La motivation pour l'étude du contrôle de systèmes à états multiples est fournie par le *contrôle de systèmes enzymatiques*, dont une étude est faite dans le livre en préparation de J.P. Kernevez, G. Thomas et J.L. Lions.

On trouvera dans ce livre l'étude d'autres situations : contrôle de systèmes évolutifs *instables*, contrôle de systèmes *périodiques*, etc ainsi que des *algorithmes numériques* (basés notament sur les travaux de MMs. Duban et Joly).

Le contrôle de systèmes *mal posés* (par exemple problème de Cauchy pour un système elliptique, ou équation de la chaleur rétrograde, etc.) a été étudié dans J.L. Lions [2] [6] et sera repris dans [5].

L'étude du contrôle de systèmes instables paraboliques et hyperboliques est faite dans J.L. Lions [3] [4] et [5].

Le contrôle optimal du système de Navier Stokes (avec une viscosité éventuellement *négative*, donc un système mal posé) a été étudié par Foursikov [1] [2] [3].

1.3. Orientation

Nous allons d'abord montrer, au N° 2, l'*existence* d'au moins un couple $\{u,y\}$ tel que

$$(1.13)\qquad \begin{cases} J(u,y) = \inf. J(v,z) \quad , \quad \{v,z\} \in E(\mathcal{U}_{ad}) \ , \\ u,y \in E(\mathcal{U}_{ad}) \ ; \end{cases}$$

on dit que $\{u,y\}$ est un *couple optimal.*

On étudie ensuite au N° 3 la structure du *système d'optimalité* donnant des conditions nécessaires satisfaites par $\{u,y\}$ couple optimal quelconque.

2. Existence

Théorème 2.1. Il existe $\{u,y\}$ *, couple optimal, vérifiant (1.13).*

Remarque 2.1.

Il n'y a aucune raison pour qu'il y ait *unicité* du couple optimal $\{u,y\}$.

Démonstration

C'est immédiat. Soit en effet $\{u_n, z_n\}$ une suite minimisante.

Il résulte de la structure de (1.5) que

(2.1) $\{v_n, z_n\}$ demeure dans un borné de $L^2(\Omega) \times L^6(\Omega)$.

Mais on a

(2.2) $-\Delta z_n = v_n + z_n^3$, $z_n = 0$ sur Γ

de sorte que

(2.3) z_n demeure dans un borné de $H^2(\Omega) \cap H_0^1(\Omega)$.

Il en résulte que l'on peut extraire une sous suite, encore notée $\{v_n, z_n\}$, telle que

(2.4) $$\begin{cases} v_n \to u & \text{dans } L^2(\Omega) \text{ faible ,} \\ z_n \to y & \text{dans } H^2(\Omega) \cap H_0^1(\Omega) \text{ faible } {}^{(1)} \end{cases}$$

et

(2.5) $z_n(x) \to y(x)$ p.p. dans Ω.

[En effet l'injection de $H_0^1(\Omega)$ dans $L^2(\Omega)$ est *compacte*; en fait, il y a bien davantage : $H^2(\Omega) \subset C^0(\bar{\Omega})$ = fonctions continues dans $\bar{\Omega}$ et on peut supposer que $z_n \to y$ uniformément dans Ω] .

Donc, on obtient à la limite :

(2.6) $$\begin{cases} u \in \mathcal{U}_{ad} , \\ - \Delta y - y^3 = u \text{ dans } \Omega, \quad y = 0 \text{ sur } \Gamma \end{cases}$$

et $\underline{\lim}\, J(v_n, z_n) \geq J(u,y)$

---

(1) Ce qui entraîne que $z_n \to y$ dans $L^6(\Omega)$ faible.

de sorte que $J(u,y) = \inf. J(v,z)$, $\{v,z\} \in E(\mathcal{U}_{ad})$, et le Théorème est démontré. □

3. Système d'optimalité

3.1. Résultat principal

On fera l'*hypothèse*

(3.1) il existe un ouvert $\omega \subset \Omega$ tel que $\mathcal{U}_{ad} \supset \tilde{\mathcal{D}}(\omega)$

(i.e. $\mathcal{U}_{ad}$ contient toutes les fonctions $\tilde{\phi}$ où ϕ est indéfiniment différentiable à support compact dans ω et où $\tilde{\phi}$ désigne le prolongement de ϕ par 0 hors de ω).

On va démontrer le

Théorème 3.1. On suppose que (3.1) a lieu ainsi que (1.10). Il existe alors un triplet $\{u,y,p\}$ *tel que* $\{u,y\}$ *soit un couple optimal et tel que les équations et inéquations suivantes soient satisfaites :*

(3.2) $u \in \mathcal{U}_{ad}$, $y \in H^2 \cap H^1_0(\Omega)$, $p \in W^{2,6/5}(\Omega)$ (1)

(3.3)
$$\left|\begin{array}{l} -\Delta y - y^3 = u , \\ -\Delta p - 3y^2 p = (y-z_d)^5 \quad \text{dans } \Omega, \\ y = p = 0 \quad \text{sur } \Gamma \end{array}\right.$$

(3.4) $(p + Nu, v - u) \geq 0 \quad \forall\, v \in \mathcal{U}_{ad}$.

Remarque 3.1.

D'après le théorème de plongement de Sobolev, on a :

(3.5)
$$\left|\begin{array}{l} W^{2,6/5}(\Omega) \subset L^6(\Omega) \quad \text{si } n = 3 , \\ W^{2,6/5}(\Omega) \subset C^0(\bar{\Omega}) \quad \text{si } n = 2 \end{array}\right.$$

---

(1) $W^{2,\beta}(\Omega) = \{\phi \,|\, \phi, \frac{\partial \phi}{\partial x_i} , \frac{\partial^2 \phi}{\partial x_i \partial x_j} \in L^\beta(\Omega) \quad \forall\, i,j\}$.

de sorte que le produit scalaire dans (3.4) *a un sens*.

Remarque 3.2.

P. Rivera [1] a démontré un résultat analogue, *sans l'hypothèse (3.1)*, mais en supposant que $||z_d||_{L^6(\Omega)}$ est *assez petit*. □

La démonstration du Théorème 3.1. s'effectue en plusieurs étapes.

3.2. Pénalisation

On introduit, pour $\varepsilon > 0$:

$$J_\varepsilon(v,z) = \frac{1}{6} ||z-z_d||^6_{L^6(\Omega)} + \frac{N}{2} ||v||^2_{L^2(\Omega)} + \frac{1}{2\varepsilon} ||\Delta z+z^3+v||^2_{L^2(\Omega)} , \tag{3.6}$$

où

$$v \in \mathcal{U}_{ad} , \tag{3.7}$$

$$z \in L^6(\Omega) \quad , \quad \Delta z \in L^2(\Omega) \quad , \quad z|_\Gamma = 0 . \tag{3.8}$$

(Donc, comme précédemment, $z \in H^2(\Omega) \cap H^1_0(\Omega)$).

On considère le *problème pénalisé* :

$$\text{inf. } J_\varepsilon(v,z) \quad , \quad v,z \text{ vérifiant (3.7) (3.8).} \tag{3.9}$$

On vérifie que

$$\left| \begin{array}{l} \text{il existe } \{u_\varepsilon,y_\varepsilon\}, \text{ avec} \\ J_\varepsilon(u_\varepsilon,y_\varepsilon) = \text{inf. } J_\varepsilon(v,z) . \end{array} \right. \tag{3.10}$$

[Comme à la démonstration du Théorème 2.1.].

On établit maintenant les premières estimations a priori et la *convergence* de $\{u_\varepsilon,y_\varepsilon\}$ (en fait d'une suite extraite) vers *un* couple optimal $\{u,y\}$.

3.3. Estimations a priori (I) et convergence

Grâce à (1.10), on a :

$$J_\varepsilon(u_\varepsilon,y_\varepsilon) \leq J_\varepsilon(u_0,z_0) = J(v_0,z_0) = \text{constante}$$

donc, lorsque $\varepsilon \to 0$,

(3.11) $\quad \{u_\varepsilon,y_\varepsilon\}$ demeure dans un borné de $L^2(\Omega) \times L^6(\Omega)$,

et

(3.12)
$$\left| \begin{array}{l} \Delta y_\varepsilon + y_\varepsilon^3 + u_\varepsilon = \sqrt{\varepsilon}\, f_\varepsilon \quad , \quad f_\varepsilon \text{ borné dans } L^2(\Omega) \ , \\ y_\varepsilon|_\Gamma = 0 \ . \end{array} \right.$$

Il résulte de (3.11) et (3.12) que

(3.13) $\quad y_\varepsilon$ demeure dans un borné de $H^2(\Omega) \cap H_0^1(\Omega)$.

Par conséquent on peut extraire une sous suite, encore notée $\{u_\varepsilon,y_\varepsilon\}$, telle que

(3.14)
$$\left| \begin{array}{l} \{u_\varepsilon,y_\varepsilon\} \to \{u,y\} \text{ dans } L^2(\Omega) \times L^6(\Omega) \text{ faible ,} \\ y_\varepsilon \to y \text{ dans } H^2(\Omega) \cap H_0^1(\Omega) \text{ faible ,} \end{array} \right.$$

(3.15) $\quad - \Delta y - y^3 = u$ dans Ω, $y|_\Gamma = 0$.

On a :

(3.16) $\quad J(u_\varepsilon,y_\varepsilon) \leq J_\varepsilon(u_\varepsilon,y_\varepsilon) \leq J_\varepsilon(v,z) \leq J(v,z)$ si $\{v,z\} \in E(\mathcal{U}_{ad})$

donc

$$J(u_\varepsilon,y_\varepsilon) \leq J_\varepsilon(u_\varepsilon,y_\varepsilon) \leq \inf J(v,z) \quad , \quad \{v,z\} \in E(\mathcal{U}_{ad})$$

et donc

(3.17) $\quad J(u,y) \leq \underline{\lim}\, J(u_\varepsilon,y_\varepsilon) \leq \overline{\lim}\, J_\varepsilon(u_\varepsilon,y_\varepsilon) \leq \inf J(v,z)$

d'où il résulte que

(3.18) $J(u,y) = \inf. J(v,z)$

et que

$$J(u_\varepsilon,y_\varepsilon) \to J(u,y).$$

Par conséquent

(3.19) $$\left| \begin{array}{l} u_\varepsilon \to u \text{ dans } L^2(\Omega) \textit{ fort} \\ y_\varepsilon \to y \text{ dans } L^6(\Omega) \textit{ fort } ^{(1)} \end{array} \right. .$$

3.4. Système d'optimalité pénalisé

On écrit maintenant la condition nécessaire d'Euler exprimant que $\{u_\varepsilon,y_\varepsilon\}$ réalise le minimum de $J_\varepsilon(u,z)$ pour $v \in \mathcal{U}_{ad}$. On pose

(3.20) $$p_\varepsilon = \frac{1}{\varepsilon}(\Delta y_\varepsilon + y_\varepsilon^3 + u_\varepsilon) .$$

On a :

(3.21) $$\left| \begin{array}{l} (p_\varepsilon, \Delta\zeta + 3y_\varepsilon^2\zeta) + ((y_\varepsilon - z_d)^5,\zeta) = 0 \\ \forall \zeta \text{ vérifiant } (3.8) \end{array} \right.$$

et

(3.22) $$(p_\varepsilon + Nu_\varepsilon , v - u_\varepsilon) \geq 0 \quad \forall v \in \mathcal{U}_{ad} .$$

Tout revient donc à établir des estimations a priori sur p_ε.

Remarque 3.3.

Naturellement c'est *immédiat* si $\mathcal{U}_{ad} = L^2(\Omega)$ — cas sans contrainte puisque dans ce cas (3.22) équivaut à

(3.23) $$p_\varepsilon + Nu_\varepsilon = 0$$

---

(1) Ce qui résulte aussi du fait que $y_\varepsilon \to y$ dans $H^2(\Omega)$ faible.

d'où, d'après (3.18)

(3.24) $\quad p_\varepsilon$ *est borné dans* $L^2(\Omega)$. □

3.5. Estimations a priori (II)

On va démontrer le

Lemme 3.1. Sous l'hypothèse (3.1), p_ε *demeure dans un borné de* $L^2(\Omega)$.

Démonstration.

On raisonne par l'absurde. Supposons que

(3.25) $$||p_\varepsilon||_{L^2(\Omega)} \to \infty .$$

On introduit :

(3.26) $$q_\varepsilon = \frac{p_\varepsilon}{||p_\varepsilon||_{L^2(\Omega)}} .$$

Il résulte de (3.21) que

(3.27) $$\begin{cases} -\Delta q_\varepsilon - 3y_\varepsilon^2\, q_\varepsilon = (y_\varepsilon - z_d)^5/||p_\varepsilon||_{L^2(\Omega)} , \\ q_\varepsilon|_\Gamma = 0 . \end{cases}$$

Mais comme $||q_\varepsilon||_{L^2(\Omega)} = 1$ on a :

$y_\varepsilon^2\, q_\varepsilon$ demeure dans un borné de $L^{6/5}(\Omega)$,

et donc (3.27) et les estimations a priori classiques dans les problèmes elliptiques entrainent que

(3.28) $\quad q_\varepsilon$ est borné dans $W^{2,6/5}(\Omega)$.

D'après (3.5) et les résultats de Sobolev, l'injection de $W^{2,6/5}(\Omega)$ dans $L^2(\Omega)$ est *compacte*. Donc, on peut extraire une sous suite, encore notée q_ε, telle que

(3.29) $q_\varepsilon \to q_0$ dans $L^2(\Omega)$ *fort* et dans $W^{2,6/5}(\Omega)$ faible,

(3.30) $$\left| \begin{array}{l} -\Delta q_0 - 3y^2 q_0 = 0 \text{ dans } \Omega, \\ q_0|_\Gamma = 0 \end{array} \right.$$

et

(3.31) $$||q_0||_{L^2(\Omega)} = 1 .$$

Mais (3.22) donne, par division par $||p_\varepsilon||_{L^2(\Omega)}$:

$$(q_\varepsilon + N \frac{u_\varepsilon}{||p_\varepsilon||_{L^2(\Omega)}} , v-u_\varepsilon) \geq 0$$

d'où, à la limite

(3.32) $$(q_0, v-u) \geq 0 \quad \forall\, v \in \mathcal{U}_{ad} .$$

Mais, utilisant l'hypothèse (3.1), (3.32) entraine que

(3.33) $$q_0 = 0 \text{ dans } \omega.$$

Il résulte de (3.30) et (3.33) et du *théorème de prolongement unique pour les équations elliptiques* (1) que

(3.34) $$q_0 = 0 ,$$

ce qui contredit (3.31), d'où le lemme.

3.6. Fin de la démonstration

On déduit de (3.21) que

(3.35) $$-\Delta p_\varepsilon - 3y_\varepsilon^2 p_\varepsilon = (y_\varepsilon - z_d)^5 \quad , \quad p_\varepsilon|_\Gamma = 0 .$$

Comme p_ε est borné dans $L^2(\Omega)$, on a : $y_\varepsilon^2 p_\varepsilon$ borné (en particulier) dans $L^{6/5}(\Omega)$ et donc (3.35) entraine que

(1) On utilise ici un résultat classique, car $y \in C^0(\bar{\Omega})$. Pour d'autres fonctions coût, il faut utiliser des versions beaucoup plus élaborées de ce résultat, pour lequel nous renvoyons à J.C. Saut et B. Scheurer [1] et à la bibliographie de ce travail. Pour les applications, cf. J.L. Lions [5].

(3.36) $\quad p_\varepsilon$ demeure dans un borné de $W^{2,6/5}(\Omega)$.

On peut donc extraire une sous suite (de la sous suite déjà extraite en (3.14)) telle que

(3.37) $\quad p_\varepsilon \to p$ dans $W^{2,6/5}(\Omega)$ faible.

Il n'y a alors aucune difficulté à passer à la limite, ce qui démontre le Théorème.

4. Remarques diverses

Remarque 4.1.

On peut étudier, par des méthodes analogues, le cas des *contrôles frontières; par exemple*

(4.1)
$$\begin{cases} -\Delta z - z^3 = f \text{ dans } \Omega, \ f \text{ donné}, \\ \dfrac{\partial z}{\partial \nu} = v \text{ sur } \Gamma \ (\text{ou } z = v \text{ sur } \Gamma) \end{cases}$$

avec

(4.2) $\quad v \in L^2(\Gamma)$,

ou

(4.3) $\quad v \in \mathcal{U}_{ad} \subset L^2(\Gamma)$.

Si

(4.4) $\quad \mathcal{U}_{ad} \supset \widetilde{\mathcal{D}}(\Gamma_0)$, Γ_0 ouvert $\subset \Gamma$,

ou bien (P. Rive ra) si $||z_d||_{L^6(\Omega)}$ est assez petit, on a encore un système d'optimalité ayant même structure générale qu'au Théorème 3.1. Cf. J.L. Lions [5].

Remarque 4.2.

Ce qui a été fait s'étend à des *systèmes* elliptiques, une situation qui est indispensable dans les applications. Cf. J.P. Kernevez, J.L. Lions et D. Thomas [1].

Remarque 4.3.

L'analyse *directe* du système (3.3) (3.4) (par exemple "combien" ce système a-t-il de solutions ?) est un problème ouvert qu'il serait intéressant d'approfondir.

Pour des algorithmes numériques basés sur des systèmes d'optimalité de ce type, nous renvoyons à M.C. Duban [1], G. Joly [1].

Bibliographie

A. BAHRI [1] Topological results on a certain class of functionals and application. J. Funct. Anal. 1982.

A. BAHRI et H. BERESTYCKI [1] A perturbation method in critical point theory and application. Trans. A.M.S. 1982.

M.C. DUBAN [1] A paraître.

A.V. FOURSIKOV [1] Problèmes de contrôle ... Mat. S. bornik. 1981, 115(157) : 2(6), p. 281-307.

[2] A propos de la résolution unique du système de Navier Stokes... Ouspechi Mat. Nauk., 1981, 36 : 2, p. 207-208.

[3] Propriétés des solutions de certains problèmes d'extremum.. Ouspechi Mat. Nauk, 1981, 36 : 5, p. 222-223.

G. JOLY [1] A paraître.

J.P. KERNEVEZ, J.L. LIONS, G. THOMAS [1] *Contrôle optimal de systèmes enzymatiques*. En préparation.

J.L. LIONS [1] *Sur le contrôle optimal des systèmes gouvernés par des équations aux dérivées partielles*. Paris, Dunod Gauthier Villars, 1968 (Traduction anglaise par S.K. Mitter, Springer, 1971).

[2] Cours Collège de France, Automne 1980 et Automne 1981.

[3] Optimal control of non well posed distributed Systems and related non linear partial differential equations. Colloque. Los Alamos. Mars 1981.

[4] On the optimal control of unstable distributed systems. Colloque Novosibirsk. Juin 1981.

[5] *Controle optimal de systèmes distribués imparfaits*. En préparation.

[6] *Some methods in the mathematical analysis of systems and their control*. Science Press. BEIJING, 1981.

P. RIVERA [1] A paraître.

J.C. SAUT et B. SCHEURER [1] Sur l'unicité du problème de Cauchy et le prolongement unique pour des équations elliptiques à coefficients non localement bornés. J. Diff. Equations, 1982.

A relation between existence of minima for non convex integrals and uniqueness for non strictly convex integrals of the calculus of variations

PAOLO MARCELLINI

Let us consider an integral of the calculus of variations of the following type

$$(1) \qquad F(v) = \int_\Omega f(x,v(x),Dv(x))\,dx$$

where Ω is an open set of $\mathbb{R}^n$, v is a real function defined in Ω with distributional first derivatives in $(L^p(\Omega))^n$ for some $p > 1$ (i.e. v is a function of the Sobolev space $H^{1,p}(\Omega)$), and $f(x,s,\xi)$ is a Caratheodory function, i.e. measurable in x and continuous in (s,ξ).

The direct method of the calculus of variations to get the existence of minima of the given integral in some subset of $H^{1,p}(\Omega)$ is based on the lower semicontinuity of F in the (sequential) weak topology of $H^{1,p}(\Omega)$. The semicontinuity of F has been well studied by many authors. We recall for example: Serrin [22], De Giorgi [7], Berkowitz [5], Cesari [6], Ioffe [11], Olech [20]. In these papers similar, but different hypotheses are considered; but in all of them the assumption that $f(x,s,\xi)$ is a convex function with respect to ξ plays a crucial role.

In fact it has been proved that convexity of f with respect to ξ is also necessary to the sequential lower semicontinuity of the given integral. This has been discovered from the very beginning by Tonelli [23], and then by Morrey [17],[18], Ioffe [11], Olech [20], Ekeland and Temam [8], Marcellini and Sbordone [14] and by Oppezzi [21].

Let us mention that if v is a vector valued function, i.e. if $v \in (H^{1,p}(\Omega))^N$ for some $N > 1$, then convexity is no more necessary for semicontinuity. Convexity must be replaced by the so called quasi-convexity, introduced by Morrey [17] in 1952. For this condition we refer also to the book of Morrey [18], and the paper of Ball [4]. Recently, under this quasi-convexity assumption, a theorem of existence of minima has been given in [15], and a semicontinuity result has been proved by Acerbi and Fusco [1].

Let us come back to scalar functions $v \in H^{1,p}(\Omega)$. We have already mentioned that convexity is necessary for semicontinuity. But, in spite of this fact, convexity is not necessary for existence of minima. To see this we consider the one-dimensional case ($n = 1$ and $\Omega = (a,b)$):

THEOREM 1. - <u>Let</u> $f(x,\xi)$ <u>be a Caratheodory function (not necessarily convex with respect to</u> ξ) <u>such that</u>

(i) $f(x,\xi) \geq \lambda|\xi|^p$ <u>for some</u> $\lambda > 0$ <u>and</u> $p > 1$;

(ii) $f(\cdot,\xi) \in L^1(a,b)$.

<u>Then there exists the minimum of the integral</u>

$$F(v) = \int_a^b f(x,v'(x))\,dx$$

<u>among the functions of</u> $H^{1,p}(a,b)$ <u>with given boundary values at the endpoints</u> $x = a$ <u>and</u> $x = b$.

When I was on the point of publishing this result in [13], I discovered that a similar theorem had been already announced, under some further assumptions such as i.e. $p=2$, by Aubert and Tahraoui in [2] (cfr. [3] for the proofs). After my lecture Jerald Goodman has pointed out to me the paper of Klotzler [12], where is considered a non convex integral that attains its infimum; and Czeslaw Olech has given me a reprint of his paper [19], where he obtained theorem 1 in the setting of control theory. It is curious that none of these authors seemed to know, on reading the references, the papers of the others. Of course the natural order of the quoted papers is [19], [12], [2], [3], [13].

Is true a result of the type of theorem 1 in n dimensions ?

We will show that the answer is negative for some special integrands and some boundary conditions (cfr. also theorem 3 of [13]). A first step in showing this is given by the proposition that follows. We assume that $f(x,s,\xi)$ is bounded for bounded s and ξ. As usual, if f is not convex with respect to ξ, by $f^{**}(x,s,\xi)$ we denote the greatest function, convex with respect to ξ, and less than or equal to f. We assume that f^{**} is a Caratheodory function; this happens for example if f is independent of s or if $f(x,s,\xi) \geqslant \lambda|\xi|^p$ for some $\lambda > 0$ and $p > 1$ (see [14]; corollary 3.12). A function $u_o \in H^{1,p}(\Omega)$ is a fixed boundary datum.

PROPOSITION 2. - <u>If</u> u <u>realizes the minimum on</u> $u_o + H_o^{1,p}(\Omega)$ <u>of the integral</u>

$$(2) \qquad \int_\Omega f(x,v,Dv)\,dx \quad ,$$

<u>then</u> u <u>realizes the minimum on</u> $u_o + H_o^{1,p}(\Omega)$ <u>also of the integral</u>

$$(3) \qquad \int_\Omega f^{**}(x,v,Dv)\,dx \quad .$$

<u>Viceversa, if the integral (3) has on</u> $u_o + H_o^{1,p}(\Omega)$ <u>a unique minimum</u> u , <u>and if</u> $f(x,u(x),Du(x)) \neq f^{**}(x,u(x),Du(x))$ <u>on a set of positive measure, then the integral in (2) does not have a minimum on</u> $u_o + H_o^{1,p}(\Omega)$.

PROOF. - We have (see [8], [14])

$$\inf \left\{ \int_\Omega f(x,v,Dv)\,dx \quad : \quad v \in u_o + H_o^{1,p}(\Omega) \right\} =$$

$$\inf \left\{ \int_\Omega f^{**}(x,v,Dv)\,dx \quad : \quad v \in u_o + H_o^{1,p}(\Omega) \right\} .$$

The first part of the proposition follows at once.

Let us denote by m the common value of the infima. If v is different from the unique minimizing function u of the integral in (3), we have

$$\int_\Omega f(x,v,Dv)\,dx \geq \int_\Omega f^{**}(x,v,Dv)\,dx > m \quad .$$

Therefore the integral in (2) does not achieve its infimum.

We have shown in the previous proposition that the existence of minima for a given integral as in (2) is related to the uniqueness of the minimizing function for the "relaxed" integral in (3). The standard argument of strict convexity to get uniqueness of the minimum is not applicable to the integral in (3), since f^{**} is not strictly convex where $f \neq f^{**}$. Here we describe a particular case in which it is possible to get uniqueness without strict convexity.

THEOREM 3. - *Let* $g : [0,+\infty) \to [0,+\infty]$ *be a convex function with* $g(0) < g(t)$ *for every* $t > 0$. *Let* Ω *be a convex bounded open set of* $\mathbb{R}^n$ $(n \geq 2)$ *with boundary* $\partial\Omega$ *of class* C^1. *If there is a function* $u \in C^1(\overline{\Omega})$, *with* $Du \neq 0$ *in* $\overline{\Omega}$, *that minimizes the integral*

$$(4) \qquad \int_\Omega g(|Dv(x)|)\, dx$$

among all Lipschitz-continuous functions that assume the same value of u *on the boundary* $\partial\Omega$, *then* u *is the unique minimizing function of this class.*

REMARK 4. - A well known theorem of Hartman and Stampacchia [10] ensures that there are minima in the class of all Lipschitz-continuous functions, if it is satisfied the "bounded slope condition" (see [10]).

In theorem 3 the assumption that Du is continuous and nonzero in $\overline{\Omega}$ seems to be technical. It should be interesting to know if it is possible to eliminate this hypothesis. On the contrary, it is crucial the assumption that the datum at the boundary is continuous, at least if we look for solutions $u \in BV(\Omega)$, the space of functions of $L^1(\Omega)$ with derivatives which are measures with bounded total variation. In fact let us consider $g(t) = t$ and let us extend the functional

$$\int_\Omega |Dv|\, dx$$

to be the total variation of Dv. The functions u_1 and u_2 defined by

$$u_1(x_1,x_2) = \pm 1 \quad \text{if} \quad |x_1| \lesseqgtr 1 \quad ,$$
$$u_2(x_1,x_2) = \pm 1 \quad \text{if} \quad |x_2| \gtreqless 1 \quad ,$$

assume the same value on the boundary of the disk $\Omega = \{(x_1,x_2) \in \mathbb{R}^2 : x_1^2 + x_2^2 < 1\}$ and both (and any convex combination of u_1 and u_2 too) realize the minimum on $BV(\Omega)$ of the given functional with the given boundary datum.

In the proof of theorem 3 we will use the result that follows. We assume the same notations and the same hypotheses of theorem 3. We assume also that u has been extended (this is certainly possible) to a C^1 function on an open bounded set $\Omega' \supset \overline{\Omega}$.

LEMMA 5. - <u>Let</u> $\Omega_\circ$ <u>be a set of</u> $\mathbb{R}^n$ <u>of zero measure. For almost every</u> $\overline{x} \in \Omega$ (<u>for every</u> $\overline{x}$, <u>if</u> $\Omega_\circ = \emptyset$) <u>there is a curve</u> $x(t)$, $t \in [a,b]$, <u>piecewise of class</u> C^1, <u>such that</u> $x(a) = \overline{x}$, $x(b) \in \partial\Omega$, $x(t) \notin \Omega_\circ$ <u>for almost every</u> $t \in [a,b]$, <u>and</u> $u(x(t))$ <u>is constant with respect to</u> t.

PROOF. - Let m and M be respectively the infimum and the supremum of $u(x)$ for $x \in \Omega'$. For $t \in (m,M)$ we consider the level sets of u, $L(t) = \{x \in \Omega' : u(x) = t\}$. Since $Du \neq 0$, $L(t)$ is of class C^1 for every t. If $\chi_{\Omega_\circ}(x)$ is the characteristic function of $\Omega_\circ$, we have

$$0 = \int_{\Omega'} \chi_{\Omega_0}(x)\,dx = \int_m^M dt \int_{L(t)} \chi_{\Omega_0}(x)\,d\sigma \quad ;$$

therefore the $(n-1)$ dimensional measure of $L(t) \cap \Omega_0$ is zero for almost every $t \in (m,M)$. It follows that for almost every $\bar{x} \in \Omega'$ the set of points of $L(u(\bar{x}))$ that are in Ω_0 has zero ($(n-1)$ dimensional) measure.

For such $\bar{x}$, we denote by M the component of $L(u(\bar{x}))$ containing $\bar{x}$ and connected. We will prove that $M \cap \partial\Omega \neq \emptyset$. To this aim, let us assume the contrary. Since $Du \neq 0$, M has in each point a normal vector that can be oriented continuously. Since M is a compact manifold, it is the boundary of an open bounded set G ($\neq \emptyset$) that, since $M \cap \partial\Omega = \emptyset$, is contained in Ω. Let us define the function $\bar{u}$ in $\bar{\Omega}$ by

$$u(x) = \begin{cases} u(x) & , \quad x \in \bar{\Omega} \setminus G \ , \\ u(\bar{x}) & , \quad x \in G \ . \end{cases}$$

The function $\bar{u}$ is Lipschitz-continuous in $\bar{\Omega}$. Since g is strictly greater than its value in zero, we have

$$\int_\Omega g(|D\bar{u}|)\,dx = \int_{\Omega\setminus G} g(|Du|)\,dx + \int_G g(0)\,dx < \int_\Omega g(|Du|)\,dx \ ,$$

and thus u is not the function that realizes the minimum of our integral. Thus the relation $M \cap \partial\Omega \neq \emptyset$ is proved.

Let us take $\bar{\bar{x}} \in M \cap \partial\Omega$ and let us denote by $A(\bar{\bar{x}})$ the set of points $x \in M$ that can be joint with $\bar{\bar{x}}$ by an arc $x(t)$, piecewise of class C^1, contained in M and such that $x(t) \notin \Omega_0$ for almost every t.

$A(\bar{\bar{x}})$ is open relatively to M; in fact, let us assume that $\bar{y} \equiv (\bar{y}_i) \in A(\bar{\bar{x}})$. Since $Du(y) \neq 0$, M can be represented locally as a graph of a function, i.e. the function

$$(5) \qquad x_n = \varphi(x_1,\ldots,x_{n-1}) \quad \text{for} \quad \sum_{i=1}^{n-1}(x_i - \bar{y}_i)^2 < \delta \quad ,$$

for some positive δ. We will show that any other point $y \equiv (y_i) \in M$, with $\sum_{i=1}^{n-1}(y_i - \bar{y}_i) < \delta$, can be joint to $\bar{y}$ by an arc with the requested properties. A consequence will be that $\bar{\bar{x}}$ can be joint to y through $\bar{y}$.

If $n = 2$, equation (5) represents a curve from $\bar{y}$ to y that solves our problem. If $n > 2$, we consider a new point $\bar{\bar{y}} \equiv (\bar{\bar{y}}_i)$ such that $\sum_{i=1}^{n-1}(\bar{\bar{y}}_i - \bar{y}_i) < \delta$ and we join y to $\bar{\bar{y}}$ by the arc $x(t) \equiv (x_i(t))$ defined for $t \in [0,1/2]$ by

$$\begin{cases} x_i(t) = y_i + 2t(\bar{\bar{y}}_i - y_i) \quad , \quad i = 1,\ldots,n-1 \; ; \\ x_n(t) = \varphi(x_1(t),\ldots,x_{n-1}(t)) \quad . \end{cases}$$

Analogously we join $\bar{\bar{y}}$ to y by an arc $x(t)$ with $t \in [1/2,1]$. Thus we have defined a curve $x(t)$, $t \in [0,1]$, piecewise of class C^1, with $x(0) = y$ and $x(1) = \bar{y}$. It is easy to see that it is possible to choose $\bar{\bar{y}}$ in such a way that $x(t) \notin \Omega_0$ a.e. for $t \in [0,1]$ (in fact we have this property for almost every $\bar{\bar{y}}$).

Thus $A(\bar{\bar{x}})$ is open. For the same reason, if we define $A(\bar{x})$ analogously to $A(\bar{\bar{x}})$, $A(\bar{x})$ is open relatively to M. Let us consider the complement of $A(\bar{x})$ and $A(\bar{\bar{x}})$, i.e. the set $B = M \setminus (A(\bar{x}) \cup A(\bar{\bar{x}}))$. Also B is open (the proof is the same as before, since every point of B can be joint by smooth archs to points of M that are neighbour). Thus we have defined two open sets $B \cup A(\bar{x})$ and $A(\bar{\bar{x}})$, whose union is M. Since M is connected, $(B \cup A(\bar{x})) \cap A(\bar{\bar{x}}) \neq \emptyset$, i.e. $A(\bar{x}) \cap A(\bar{\bar{x}}) \neq \emptyset$. If $y \in A(\bar{x}) \cap A(\bar{\bar{x}})$, then $\bar{x}$ can be joint to $\bar{\bar{x}}$ through y. This concludes the proof.

PROOF OF THEOREM 3. - Let u and v be minima of the integral in (4) and let m be the minimum value. By the convexity of the integrand we have

$$m \leq \int_{\Omega} g\left(\left|\frac{Du + Dv}{2}\right|\right) dx \leq \int_{\Omega} \frac{1}{2}\left\{g(|Du|) + g(|Dv|)\right\} dx = m ,$$

and thus

$$g\left(\frac{|Du + Dv|}{2}\right) = \frac{1}{2}\left\{g(|Du|) + g(|Dv|)\right\} \quad \text{a.e. in } \Omega .$$

This implies that g is affine in the above arguments, i.e. there exist $m(x)$ and $q(x)$ such that a.e. in Ω

$$(6) \qquad \frac{m(x)}{2}|Du + Dv| + q(x) = \frac{1}{2}\left\{m(x)|Du| + q(x) + m(x)|Dv| + q(x\right.$$

Now, since the convexity of g, and since $g(|Du|) > g(0)$, we have the slope $m(x)$ of $g(t)$ at $t = |Du|$ strictly positive. Therefore (6) implies that Du and Dv are linearly dependent and, again since $Du \neq 0$, there exists a bounded measurable function $\lambda(x)$ such that

$$(7) \qquad Dv(x) = \lambda(x)\, Du(x) \qquad \text{a.e. in } \Omega .$$

We have already extended u to an open set $\Omega' \supset \bar{\Omega}$;

we define $v = u$ on $\Omega' \setminus \Omega$; of course (7) holds in $\Omega' \setminus \Omega$ too (with $\lambda(x) = 1$).

Let us consider $v_\varepsilon = v * \alpha_\varepsilon$ where α_ε is a mollifier. There is a sequence (that we still denote by ε) going to zero such that $Dv_\varepsilon(x)$ converges to $Dv(x)$ a.e. in Ω. Let Ω_0 be the set of zero measure such that $Dv_\varepsilon(x)$ converges to $Dv(x)$ for every $x \in \Omega \setminus \Omega_0$. Now we apply lemma 5 with this Ω_0. For almost every $\bar{x} \in \Omega$ we have a piecewise-C^1 curve $x(t)$, $t \in [a,b]$, $x(a) = \bar{x}$, $x(b) \in \partial\Omega$, and $x(t) \notin \Omega_0$ a.e. in $[a,b]$. We use the relation

$$v_\varepsilon(\bar{x}) - v_\varepsilon(x(b)) = \int_b^a \sum_i D_i v_\varepsilon(x(t))\, x_i'(t)\, dt \quad .$$

We can go to the limit, as $\varepsilon \longrightarrow 0$, in the left side since v is continuous, and in the right side, since Dv is bounded. From (7) and the fact that $u(x(t))$ is constant, we obtain

$$\begin{aligned} v(\bar{x}) - v(x(b)) &= \int_b^a \sum_i D_i v(x(t))\, x_i'(t)\, dt \\ &= \int_b^a \lambda(x) \sum_i D_i u(x(t))\, x_i'(t)\, dt \\ &= \int_b^a \lambda(x) \frac{d}{dt} u(x(t))\, dt = 0 \; . \end{aligned}$$

Therefore $v(\bar{x}) = v(x(b))$ and this boundary value is the same as $v(x(b)) = u(x(b)) = u(\bar{x})$. Thus $v = u$ a.e. in Ω.

Using the previous result we can describe existence or nonexistence of minima for integrals of the type of (4) when the boundary datum is linear. To this aim we denote by $g : \mathbb{R} \to [0,+\infty]$ an even continuous function and by g^{**} the greatest convex function less than or equal to g in $\mathbb{R}$. We denote also by $(g^{**})'_+(t)$ the right derivative of g^{**} at $t \geq 0$.

THEOREM 6. - _If_ $g(t_\circ) > g^{**}(t_\circ)$ _and_ $(g^{**})'_+(t_\circ) > 0$ _for some_ $t_\circ > 0$, _then the problem_

$$\min\left\{\int_\Omega g(|Dv|)\,dx \ : \ v \ \text{is Lipschitz-cont.}, \ v = \sum_i \xi_i x_i \ \text{on}\ \partial\Omega\right\}$$

where $\xi \equiv (\xi_i)$ _is a vector of_ $\mathbb{R}^n$ _of modulus equal to_ $t_\circ$, _lacks a solution_.

THEOREM 7. - _If_ $g(0) > g^{**}(0)$ _and_ $(g^{**})'_+(0) = 0$, _and if_ $\liminf_{t \to +\infty} g(t) > g^{**}(0)$, _then the problem_

$$\min\left\{\int_\Omega g(|Dv|)\,dx \ : \ v \ \text{is Lipschitz-cont.}, \ v = 0 \ \text{on}\ \partial\Omega\right\},$$

does have a solution.

PROOF OF THEOREM 6. - Let us define

$$f(\eta) = (g^{**})'_+(t_\circ)(|\eta| - t_\circ) + g^{**}(t_\circ) .$$

From Jensen's inequality, for every v equal to $u(x) = \sum_i \xi_i x_i$ on $\partial\Omega$, we have

$$\int_\Omega f(Dv)\,dx \geq \operatorname{mis}\Omega \ f\left(\frac{1}{\operatorname{mis}\Omega}\int_\Omega Dv\,dx\right) = \operatorname{mis}\Omega \ f(\xi) = \int_\Omega f(Du)\,dx$$

Therefore $u(x)=\sum_i \xi_i x_i$ minimize the integral of the left side. From theorem 3 (by adding a constant to f we can have a positive integrand) we obtain that u is the unique minimizing function in the class of Lipschitz-continuous functions. Since $g^{**}(|\eta|) \geqslant f(\eta)$ and the equality for $\eta = \xi$, $u(x)$ is also the unique minimizing function of the integral

$$\int_\Omega g^{**}(|Dv|)\,dx \quad .$$

Now the conclusion follows from proposition 2 (where, instead of considering $u_o + H_o^{1,p}(\Omega)$, we take the infimum in the class of Lipschitz-continuous functions).

PROOF OF THEOREM 7. - We can deduce from the assumptions that there exists $t_o > 0$ such that $g(t) > g^{**}(t)$ for every $t \in (-t_o, t_o)$ and $g(t_o) = g^{**}(t_o)$. We will show that there exists a Lipschitz-continuous function $u(x)$ such that

$$(8) \qquad \begin{cases} |Du(x)| = t_o & \text{a.e. } x \in \Omega \, , \\ u(x) = 0 & x \in \partial\Omega \, . \end{cases}$$

This function u solves our problem, since

$$\int_\Omega g(|Dv|)\,dx \geqslant \int_\Omega g^{**}(|Dv|)\,dx \geqslant \text{meas}\,\Omega\; g^{**}(0)$$
$$= \text{meas}\,\Omega\; g^{**}(t_o) = \text{meas}\,\Omega\; g(t_o) = \int_\Omega g(|Du|)\,dx \quad ;$$

we have used the fact that $g^{**}(t)$ is constant for $t \in [-t_o, t_o]$. Now we show that the function

$$u(x) = t_\circ \operatorname{dist}(x, \partial\Omega)$$

is a solution of (8) (Giorgio Talenti has pointed out to me lemma 3.2.34 of Federer [9], where is computed $|Du|$). It is clear that $u = 0$ on $\partial\Omega$ and that $|Du| \leqslant t_\circ$, since u is Lipschitz-continuous with constant $t_\circ$. Let us fix a point x where u is differentiable. If $\operatorname{dist}(x, \partial\Omega) =$ $= |x - y|$ for some $y \in \partial\Omega$, then on the line $x(t) = y + t(x - y)$, $t \in [0,1]$, we have

$$\operatorname{dist}(x(t), \partial\Omega) = |x(t) - y| = t|x - y| .$$

Let us denote by $\tau = (x - y)|x - y|^{-1}$ the unit vector in the direction of $x - y$. We have

$$\frac{\partial}{\partial\tau} u(x) = \frac{1}{|x - y|} \frac{d}{dt} u(x(t)) = t_\circ .$$

This implies that $|Du(x)| = t_\circ$. Since u is differentiable almost everywhere in Ω, u solves (8) and our proof is complete.

Why the situation of theorem 6 is so different from the other one of theorem 7 ?

A difference is in the structure of the function $f^{**}(\xi) = g^{**}(|\xi|)$ on the set

$$\text{(9)} \qquad K = \left\{ \xi \in \mathbb{R}^n : g(|\xi|) > g^{**}(|\xi|) \right\} .$$

In the case of theorem 7 K is a ball and $f^{**}(\xi)$ is affine (in particular constant) on K. In the other case

K is an n-dimensional circular crown; $g^{**}(t)$ is also affine, but $f^{**}(\xi)$ is not affine on K. This difference is important; we note that when f^{**} is affine on K, then the integral of $f^{**}(Dv)$ depends only on the values of v at the boundary, provided that Dv belongs a.e. to K. Thus the assumption of uniqueness in proposition 2 (that guarantees the nonexistence of minimum for the initial problem) is not satisfied.

The sufficiency for existence of f^{**} affine, has been pointed out by Mascolo and Schianchi [16]. They prove that, if f^{**} is affine on the set K where $f^{**}(\xi) < f(\xi)$, then the Dirichlet minimization problem for some boundary conditions does have a solution. The pro of is based on solving a first order problem of the type of (8).

REFERENCES

[1] E.ACERBI - N.FUSCO, Semicontinuity problems in the calculus of variations, Arch. Rat. Mech. Analysis, to appear.

[2] G.AUBERT - R.TAHRAOUI, Théorèmes d'existence en calcul des variations, C. R. Acad. Sc. Paris, 285 (1977), 355-356.

[3] G.AUBERT - R.TAHRAOUI, Théorèmes d'existence pour des problèmes du calcul des variations..., J. Differential Equations, 33 (1979), 1-15.

[4] J.M.BALL, Convexity conditions and existence theorems in nonlinear elasticity, Arch. Rat. Mech. Analysis, 63 (1977), 337-403.

[5] L.D.BERKOWITZ, Lower semicontinuity of integral functionals, Trans. Am. Math. Soc., 192 (1974), 51-57.

[6] L.CESARI, Lower semicontinuity and lower closure theorems without seminormality condition, Annali Mat. Pura Appl., 98 (1974), 381-397.

[7] E.DE GIORGI, Teoremi di semicontinuità nel calcolo delle variazioni, Istit. Naz. Alta Mat., Roma (1968-1969).

[8] I.EKELAND - R.TEMAM, Analyse convexe et problèmes variationnels, Dunod Gauthier-Villars, 1974.

[9] H.FEDERER, Geometric measure theory, Die Grundl. Math. Wiss. 153, Springer-Verlag, 1969.

[10] P.HARTMAN - G.STAMPACCHIA, On some non-linear elliptic differential-functional equations, Acta Math., 115 (1966), 271-310.

[11] A.D.IOFFE, On lower semicontinuity of integral functional I, SIAM J. Cont. Optimization, 15 (1977), 521-538.

[12] R.KLOTZLER, On the existence of optimal processes, Banach Center Publications, Volume 1, Warszawa 1976, 125-130.

[13] P. MARCELLINI, Alcune osservazioni sull'esistenza del minimo di integrali del calcolo delle variazioni senza ipotesi di convessità, Rendiconti Mat., 13 (1980), 271-281.

[14] P.MARCELLINI - C.SBORDONE, Semicontinuity problems in the calculus of variations, Nonlinear Analysis, 4 (1980), 241-257.

[15] P.MARCELLINI - C.SBORDONE, On the existence of minima of multiple integrals of the calculus of variations, J. Math. Pures Appl., to appear.

[16] E.MASCOLO - R.SCHIANCHI, Existence theorems for non convex problems, to appear.

[17] C.B.MORREY, Quasiconvexity and the lower semicontinuity of multiple integrals, Pacific J. Math., 2 (1952), 25-53.

[18] C.B.MORREY, Multiple integrals in the calculus of variations, Die Grundl. Math. Wiss. 130, Springer-Verlag, 1966.

[19] C.OLECH, Integrals of set-valued functions and linear optimal control problems, Colloque sur la Théorie Mathématique du Contrôle Optimal, C.B.R.M., Vander Louvain (1970), 109-125.

[20] C.OLECH, A characterization of L^1-weak lower semicontinuity of integral functional, Bull. Acad. Pol. Sci. Ser. Sci. Math. Astronom. Phys., 25 (1977), 135-142.

[21] P.OPPEZZI, Convessità della integranda in un funzionale del calcolo delle variazioni, Boll. Un. Mat. Ital., to appear.

[22] J.SERRIN, On the definition and properties of certain variational integrals, Trans. Am. Math. Soc., 101 (1961), 139-167.

[23] L.TONELLI, Fondamenti di calcolo delle variazioni, Volume I, Zanichelli, 1921.

Remarks on Pathwise Nonlinear Filtering

by

Sanjoy K. Mitter

Department of Electrical Engineering
and Computer Science
MASSACHUSETTS INSTITUTE OF TECHNOLOGY

CAMBRIDGE, MASS. 02139.

This research has been supported by the Air Force Office of Scientific Research under Grants AF-AFOSR-77-3281B and 82-0135.

1. Introduction

This paper is concerned with an example of a nonlinear filtering problem where it is not known whether the pathwise equations of nonlinear filtering can be used to construct the unnormalized conditional measure. For details about pathwise nonlinear filtering see CLARK [1978]

2. The Example

Consider the nonlinear filtering problem

$$(1) \qquad \begin{cases} dx_1(t) = dw_1(t) \\ dx_2(t) = dw_2(t) \end{cases} \qquad \underline{\text{state equation}}$$

$$(2) \qquad dy(t) = \left(x_1^3(t) + x_2^3(t)\right) dt + d\eta(t) \qquad \underline{\text{observation equation}}$$

where it is assumed that w_i, w_2 and η are independent Browian motions.

If $\rho(t,x)$ denotes the unnormalized conditional density of $x(t) = \left(x_1(t), x_2(t)\right)$ given $\mathscr{F}_t^y = \sigma\left\{y(s) \mid o \leq s \leq t\right\}$ then ρ satisfies the Zakai equation

$$(3) \qquad d\rho(t,x) = \frac{1}{2}\left(\Delta - (x_1^3 + x_2^3)^2\right) \rho(t,x)\, dt + (x_1^3 + x_2^3)\rho(t,x) \circ dy(t), \quad \rho(o,x) = \rho_o(x),$$

where $\circ$ denotes the Stratonovich differential, and Δ is the two-dimensional Laplacian.

Defining

(4) $\rho(t,x) = \exp\left((x_1^3 + x_2^3)\, y(t)\right) q(t,x)$, $q(t,x)$ satisfies the parabolic partial differential equation, the so-called pathwise filtering equation

(5) $$\begin{cases} q_t = \frac{1}{2}\Delta q + g^y(x,t) \cdot q_x + V^y(x,t) q \\ q(o,x) = \rho^o(x) \end{cases}$$

where $g^y(x,t) = \begin{pmatrix} y(t)\, 3x_1^2 \\ y(t)\, 3x_2^2 \end{pmatrix}$

and

$$V^y = -\frac{1}{2}(x_1^3 + x_2^3)^2 + \frac{1}{2} y^2(t)\, (9x_1^4 + 9x_2^4)$$

The difficulty with studying existence and uniqueness of solutions to (5) is that V^y is not bounded above along the direction $x_1 = -x_2$. In the corresponding scalar case, the conditional measure has been constructed using the pathwise equations by FLEMING-MITTER [1982] and SUSSMANN [1981].

3. Existence and Uniqueness of Weak Solutions

We consider the equation

(6) $$\begin{cases} \frac{du}{dt} + Au = o \\ u(o) = u_o \end{cases}$$

where $A = -\Delta + V(x_1,x_2)$, with

$$V = V_1 - V_2 = (x_1^3 + x_2^3)^2 - (x_1^4 + x_2^4) .$$

The same techniques will work for the slightly more general equation (5). The notation and terminology to be used is that of LIONS-MAGENES [1968] (Vol. 1, Chapter 3).

We define the bilinear form

$$a(\phi,\psi) = \int_{\mathbb{R}^2}\left[\frac{1}{2}\nabla\phi\cdot\nabla\psi + V(x)\phi\cdot\psi\right]dx$$

and the spaces

$$H_V^1(\mathbb{R}^2) = \left\{\phi : \mathbb{R}^2 \to \mathbb{R} \;\middle|\; D_{x_1}\phi \in L^2(\mathbb{R}^2),\ D_{x_2}\phi \in L^2(\mathbb{R}^2) \text{ and } \int_{\mathbb{R}^2}\left(V_1(x) + V_2(x)\right)|\phi(x)|^2 dx < \infty\right\}$$

with norm

$$\|\phi\|^2_{H_V^1} = \|D_{x_1}\phi\|^2_{L^2} + \|D_{x_2}\phi\|^2_{L^2} + \int_{\mathbb{R}} \Big(V_1(x) + V_2(x)\Big)|\phi(x)|^2 dx$$

and the corresponding scalar product.

$$L_V^2\ (\mathbb{R}^2) = \left\{\phi: \mathbb{R}^2 \to \mathbb{R} \ \Big|\ \int_{\mathbb{R}^2} \Big(V_1(x) + V_2(x)\Big)|\ \phi(x)\ |^2 dx < \infty\right\}$$

with the corresponding natural norm and scalar product.

It can be checked that H_V^1 and L_V^2 are complete with respect to their respective norms.

Denote by

$$\mathscr{H} = L_V^2\ (\mathbb{R}^2) \quad \text{and} \quad \mathscr{V} = H_V^1(\mathbb{R}^2)\ .$$

It is easy to check that $a(\phi,\psi)$ is a continuous bilinear form on $\mathscr{V}$ and furthermore there exists a λ such that

$a(\phi,\phi) + \lambda\ \|\phi\|^2_{\mathscr{H}} \geq \alpha\ \|\phi\|^2_{\mathscr{V}}$, $\alpha > o$. Hence by the variational theory of Parabolic equations thate exists a unique solution to the equation

$$(7) \qquad \begin{cases} \mathscr{A}u + \dfrac{du}{dt} = o \\ u(o) = u_o\ \varepsilon\ \mathscr{H}\ , \quad \mathscr{A}\ \varepsilon\ \mathscr{L}(\mathscr{V},\mathscr{V}') \end{cases}$$

in the space

$$W(o,T) = \left\{\phi\ \Big|\ \phi\,\varepsilon\,L^2\ (o,T;\mathscr{V}),\ \frac{d\phi}{dt}\ \varepsilon\ L^2(o,T;\mathscr{V}')\right\}$$

Furthermore $-\Delta + V$ generates an analytic semigroup on $\mathscr{H}$ and also using a standard regularity result the equation (7) has a C^∞-solution. It is however an <u>open problem</u> whether we have the probabilistic representatic (Feyman-Kac formula)

$$u(t,x) = E_x\left[u_o\Big(x(t)\Big)\exp\Big(\int_o^t - V\ \Big(x(s)\Big)\,ds\Big)\right]$$

where E_x denotes expectation with respect to 2-dimensional Browian motion. This case is not covered by the best results known in the Feyman-Kac formula (cf. SIMON [1979], p. 262).

Without a probabilistic representation as above it is unclear whether the conditional measure of $x(t)$ given $\mathscr{F}_t^Y$ can be constructed using the pathwise filtering equations.

Acknowledgment

I am indebted to Michel Delfour, University of Montreal for usefùl conversations about this problem.

References

Clark, J.M.C. [1978], The Design of Robust Approximations to the Stochastic Differential Equations of Nonlinear Filtering, in Communication Systems and Random Process Theory: ed. J.K. Skwirzynski, Sithoff and Noordhoff.

Fleming, W.H., and Mitter, S.K. [1982], Optimal Control and Nonlinear Filtering for Nondegenerate Diffusion Processes, to appear Stochastics.

Lions, J.L., and Magenes, E. [1968], Problèmes aux limites non homogènes et applications, Vol. 1., Dunod, Paris.

Simon, B. [1979], Functional Integration and Quantum Physics, Academic Press, New York.

Sussman, H. [1981], Rigorous Results on the Cubic Sensor Problem in Nonlinear Filtering and Stochastic Mechanics in Stochastic Systems: The Mathematics of Filtering and Identification and Applications: eds. M. Hazewinkel and J.C. Willems, pp. 479-503, D. Reidel Publishing Company.

Boundary solutions of differential inclusion

by Czesław Olech

Introduction. Consider a differential inclusion

$$(1) \qquad \dot{x} \in F(t,x), \; x \in \mathbb{R}^n$$

$F(t,x) \subset R^n$ is compact continuous in t,x and satisfies Lipschitz condition in x with respect to the Hausdorff metric.
Let us fix initial condition $x(0) = 0$ and put

$$\mathcal{A}_{(1)}(t) = \{x(t) \mid x(\cdot) \text{ a solution of } (1),\; x(0) = 0\}$$

A solution $x(t)$ of (1) is a boundary solution if $x(t) \in \partial\mathcal{A}_{(1)}(t)$ on an interval $[0,t_o]$. If Lipschitz condition holds for F in x then $x(t_o) \in \partial\mathcal{A}_{(1)}(t_o)$ implies that $x(t) \in \partial\mathcal{A}_{(1)}(t)$ for $0 < t_o$ (see A. Pliś [5]). The necessary conditions for $x(t)$ to be a boundary solution on $[0,t_o]$ is the existence of absolutly continuous $p(t)$ never equal zero on $[0,t_o]$ such that the Pontriagin maximum principle holds (see F.Clarke [1]).

$$(2) \qquad \langle \dot{x}(t), p(t) \rangle = s(t,x(t), p(t)),$$

where $s(t,x,p)$ is the support function for $F(t,x)$; that is

$$s(t,x,p) = \max_{w \in F(t,x)} \langle w,p \rangle .$$

Let $A(t)$ be an integrable $n \times n$ matrix valued function such that

$$(3) \qquad \dot{p}(t) = -p(t)A(t) \qquad t \in [0,t_o].$$

Such A exists. Consider a differential inclusion

$$(4) \qquad \dot{x} \in A(t)(x - x(t)) + F(t,x(t))$$

We call inclusion (4) a linear differential inclusion associated with (1). Notice that $x(t)$ is a solution of (4) and that (2) and (3) are necessary and sufficient conditions for $x(t)$ to be a boundary solution of (4) (see for example [4]).

Therefore we may say that $x(t)$ is a boundary solution of (1) only if there is a linear differential inclusion (4) associated with (1) such that $x(t)$ is a boundary solution of (4) also.

We discuss in this note the opposite question. We assume that $x(t)$ is a boundary solution of (4). What are the conditions for F which will imply that $x(t)$ is also a boundary solution of (1). Such conditions are given in the next section and in the section 3 the main result is stated with some indication of the proof. This note is based on two papers [2,3] of H.Frankowska and the author. There the reader can find some more details.

2. Assumptions. In the conditions which follows $x(t)$ is a solution of (4) and $p(t) \neq 0$ satisfies (3) and is such that (2) holds. To fix the ideas we assume that $t \in [0,1]$ and $|p(1)| = 1$.

The basic assumption for the result in the next section are the following two conditions:

(i) $F(t,x) \subset A(t)(x-x(t)) + F(t,x(t)) + kC(|x-x(t)|, p(t))$

(ii) $\mathcal{A}_{(4)}(1) \subset (x(1) - p(1)R, R)$

Above $B(a,r)$ denotes the closed ball of radius r and centered at a, k is a positive constant and $C(r,p) = B(0,r)$ $\{y | \langle y,p\rangle \leq r^2\}$; that is $C(r,p)$ is the intersection of the ball of radius r with the half space normal to p containing zero and the distance of zero from the hyperplace determining the half space is of order r^2.

Condition (i) says that $A(t)(x-x(t)) + F(t,x(t))$ approximates $F(t,x)$, if $x-x(t)$ is small, in the direction $p(t)$. Indeed, (i) implies the following estimate for the support function of F

$$s(t,x,p) \leq s(t,x(t),p) + pA(t)(x-x(t)) + |p-p(t)|M|x-x(t)| + k|x-x(t)|^2.$$

This inequality, in particular, implies that

$$\frac{\partial s}{\partial x}(t,x(t),\, p(t)) = p(t)A(t)$$

provided that $\frac{\partial s}{\partial x}$ exists.

On the other hand if $\frac{\partial s}{\partial x}(t,x,p)$ exists and is Lipschitz in both variables x,p then (i) holds, if in the right-hand side of (i) $F(t,x(t))$ is replaced by the convex hull $\mathrm{co}\, F(t,x(t))$ (see [3]).

Condition (ii) corresponds to the Legendre condition. In a sense it says that the second variation is positive, but it is expressed geometrically.

The important consequence of (ii) is the following estimate

$$(5) \qquad \int_0^1 |\dot{y}(t)-\dot{x}(t)|dt \leq L\left(\int_0^1 < \dot{x}(t) - \dot{y}(t),\ p(1) > dt\right)^{1/2},$$

where $y(t)$ is any solution of (4) with zero initial condition. We notice that (ii) is equivalent to the inequality

$$\left|\int_0^1 (\dot{y}(t) - \dot{x}(t))dt\right| = |y(1)-x(1)| \leq (2R<x(1)-y(1),\ p(1)>)^{1/2}$$

Thus (5) means that similar inequality holds if the norm of the integral is replaced by the integral of the norm, (see [2]).

3. The main result. We have the following

Theorem. Assume that $x(t)$ is a solution of (1) on $[0,1]$ and that there is $A(t)$ integrable such that (2) and (3) holds with $p(1) \neq 0$. Assume that both (i) and (ii) are satisfied. Under those assumptions, there is $t_o > 0$ such that $x(t)$ is a boundary solution of (1) for $0 \leq t \leq t_o$.

We notice that assumptions of the theorem are invariant with respect to linear change of variables in the state space. Thus without any loss of generality we may assume that $A(t) \equiv 0$ and $x(t) \equiv 0$. Hence $p(t) \equiv p_o = \text{const.}$

Let $y(t)$ be an arbitrary solution of (1) with the initial condition $y(0) = 0$. From (i), to each such solution, there is $v(t) \in F(t,0)$ such that

$$(7) \qquad |\dot{y}(t) - v(t)| \leq k|y(t)|$$

and

$$(8) \qquad < p, \dot{y}(t) - v(t) > \leq k|y(t)|^2$$

while condition (ii) gives the inequality

$$(9) \qquad \left(\int_0^1 |v(t)|dt\right)^2 \leq L \int_0^1 <-p, v(t)> dt$$

From inequality (7) we can estimate the norm of $y(t)$ and we get

$$|y(t)| \leq M \int_0^t |v(t)|dt$$

Using this inequality, (8) and (9) we get

$$\varphi(t) \leq < -p, y(t) > + KM^2L \int_0^t \varphi(\tau)d\tau,$$

where $\varphi(t) = \int_0^t < -p, v(t) > dt$. Which implies that there is t_o such that $<-p, y(t)> \geq 0$ for $0 \leq t < t_o$, which proves the theorem.

References

[1] Clarke, F.H., Nonsmooth Analysis and Optimization, Proceedings of the I.C.M.-78, Helsinki 1980, vol.2, 847-853.

[2] Frankowska, H. and Olech, C., R-Convexity of the integral of set-valued functions, to appear in Americ.J.of Math.

[3] Frankowska, H. and Olech, C., Boundary solution of differential inclusion, to appear in J.D.E.

[4] Hermes, H. and LaSalle, J.P., Functional Analysis and Time Optimal Control, Academic Press, New York and London 1969.

[5] Pliś, A., On trajectories of orientor fields, Bull.Acad.Polon. Sci., serie des sci. math.,astr. 13(1965) ,571-573.

Institute of Mathematics
Polish Academy of Sciences
00-950 Warszawa, P.O.Box 137
Poland

CARLO SBORDONE

On the compactness of minimizing sequences of variational problems

INTRODUCTION

Let us consider a real functional defined on a Sobolev space $(H^{1,p}(\Omega))^N$, $p \geq 1$, $N \geq 1$ integer, Ω bounded open set in R^n and satisfying

(1) $$0 \leq F(u) \leq \int_\Omega (a(x) + b|u|^p + c|Du|^p)$$

where $a(x)$, b, $c \geq 0$ and $a \in L^1(\Omega)$.

Let us fix $u_o \in (H^{1,\infty}(\Omega))^N$ and set $V = u_o + (H_o^{1,p}(\Omega))^N$.

In order to solve a variational problem

(P) to find $u \in V$ such that $F(u) = \operatorname{Min}_V F$

the direct method of the classical Calculus of Variations suggests us the following:

(A) to find a topology τ on V such that

(A_1) F is τ-sequentially lower semicontinuous

(A_2) Every minimizing sequence (u_h) of F on V has a subsequence τ-converging

So it is clear the interest of sequential lower semicontinuity results with respect to topologies like the weak topology of $(H^{1,p}(\Omega))^N$ or the strong topology of $(L^q(\Omega))^N$ $(1 \leq q \leq \infty)$, and in the first section we shall give some recent results in this direction.

The second step for solving (P) is the

(B) regularization of the solution,

in section 2 we shall describe some regularity results and prove a Liouville theorem for local minima of functionals $F = F(A,u)$ not necessarily measures with respect to A and in absence of Euler equations, generalizing some recent results of [2], [10], [17].

But here we want to report another point of view, which is:

(C) to work with (A) and (B) at the same time for finding a solution of (P).

In fact if F does not verify (A_1) and there exists a topology τ' finer than τ such that:

(A_2') there <u>exists at least one</u> minimizing sequence (u_h) of F on V which τ'-converges,

then, may be F is τ'-sequentially lower semicontinuous, and so we find a solution of (P) ([17]).

In other words we base our method on the regularization of minimizing sequences which gives a compactness condition "better" than the one which would follow directly by the coerciveness structure of the functional itself (see section 3).

1. LOWER SEMICONTINUITY THEOREMS

In the recent years a new impulsion has been given to the Calculus of Variations expecially under the initiative of E. DE GIORGI in connection with the theory of Γ-convergence [14],[15].

In particular many results have been proved about the representation of the τ-sequentially lower semicontinuous envelope $sc^-(\tau)F$ of a functional

$$F(u) = \int_\Omega f(x,Du) \qquad u\in(H^{1,p}(\Omega))^N \tag{1.1}$$

when τ is the weak topology of $(H^{1,p}(\Omega))^N$ or the strong topology of $(L^p(\Omega))^N$.

In the case N=1 we present the following two theorems.

THEOREM 1.1 - ([8]) *Let Ω be open in R^n, and* $f:\Omega \times R^n \longrightarrow R$ *verify*

$$0 \le f(x,z) \le a(x) + c|z|^p \tag{1.2}$$

where $a\in L^1(\Omega)$, $c > 0$, $p \ge 1$.

Then there exists $\bar f = \bar f(x,z) \le f^{**}(x,z) \le f(x,z)$ [(1)] *such that* $\bar f(x,\cdot)$ *is convex and*

$$sc^-(L^p(\Omega)) \int_\Omega f(x,Du) = \int_\Omega \bar f(x,Du) \qquad \forall u\in H^{1,p}(\Omega).$$

THEOREM 1.2 - ([26]) *Let* $f:\Omega\times R\times R^n \longrightarrow R$ *verify for* $x, z\in R^n, y\in R$

$$0 \le f(x,y,z) \le a(x) + b|y|^p + c|z|^p$$

---

(1) By $f^{**}(x,\cdot)$ we denote the convex envelope of $f(x,\cdot)$, that is the bipolar of $f(x,\cdot)$ in the sense of convex analysis.

where $a \in L^1(\Omega)$, $b,c>0$, $p \geq 1$ *and* $f(x,\cdot,z)$ *is continuous uniformly as* z *varies over compact sets of* R^n. *Then*

$$sc^-(w\text{-}H^{1,p}(\Omega)) \int_\Omega f(x,u,Du) = \int_\Omega f^{**}(x,u,Du)$$

for every $u \in H^{1,p}(\Omega)$, *where* $f^{**}(x,y,\cdot)$ *is the bipolar of* $f(x,y,\cdot)$.

Let us remark that while in theo. 1.2 (which, in the case $f = f(x,z)$, was proved in [16] with a different method) it is clear how to compute the new integrand f^{**}, in theo. 1.1 the expression of $\bar{f}$ is not explicit.

Let us present a recent result, relative to the case

$$f(x,z) = \sum_{i,j=1}^{n} a_{ij}(x) z_i z_j . \tag{1.3}$$

THEOREM 1.3 - ([18]) *If* f *satisfies* (1.2),(1.3), *then the integrand* $\bar{f}$ *of theo.* 1.1 *is given for a.e.* $x_0 \in \Omega$ *and* $\forall\, z \in R^n$ *by*

$$\bar{f}(x_0,z) = \sum_{i,j=1}^{n} a_{ij}(x_0) z_i z_j =$$

$$= \lim_{r\to 0}{}'' \operatorname*{Inf}_{u \in \langle z,x\rangle + H_0^1(B_r)} \frac{1}{|B_r|} \int_{B_r} f(x,Du)$$

where $B_r = B_r(x_0) = \{x : |x-x_0| < r\}$.

Let us now consider the general case $N \geq 1$ and the weak topology of $(H^{1,p}(\Omega))^N$.

There are many results in the literature about the w-$(H^{1,p}(\Omega))^N$ sequentially lower semicontinuity of (1.1); we recall the work of MORREY [28] (who introduced the notion of quasiconvexity), SERRIN [33] the work of DE GIORGI [13], BERKOWITZ [5], CESARI [9] IOFFE [24], OLECH [29], who considered as a tool the sequentially lower semicontinuity of functionals like:

$$(u,v)\in L^p(\Omega) \times (L^q(\Omega))^N \longrightarrow \int_\Omega f(x,u,v)$$

in the product topology of $s\text{-}L^p(\Omega)$ and $w\text{-}(L^q(\Omega))^N$.

This approach has been recently undertaken by BALL [4], for the study of polyconvex case.

But a very general result has been proved in [1]; before describing it , let us give MORREY's definition of quasiconvexity.

DEFINITION 1.1 - Let $f:\Omega \times R^{nN} \longrightarrow R$ be a Caratheodory function; then it is called <u>quasiconvex</u> iff $\forall\ x_o \in \Omega$, $\forall z \in R^{nN}$, $\forall w \in (C^1_o(\Omega))^N$:

$$f(x_o,z) \leq \frac{1}{|\Omega|}\int_\Omega f(x_o,z+Dw(x))dx.$$

THEOREM 1.4 ([1]) *Let* f *verify for* $(x,z)\in\Omega \times R^{nN}$

$$0 \leq f(x,z) \leq s(1+|z|^p). \tag{1.4}$$

Then

$$F(u) = \int_\Omega f(x,Du) \qquad u\in(H^{1,p}(\Omega))^N$$

is $w\text{-}(H^{1,p}(\Omega))^N$ *sequentially lower semicontinuous iff* f *is quasiconvex*.

Let us mention an interesting particular case of this result which is relevant in nonlinear elasticity:

COROLLARY 1.1 *If* $p=N=n$, $0\leq a(x)\leq s$, *then the functional*

$$F(u) = \int_\Omega a(x)\,|\det Du|\,dx$$

is $w\text{-}(H^{1,n}(\Omega))^n$ *sequentially lower semicontinuous*.

Finally we want to quote another result of [1], which generalizes theo. 1.2 to the vectorial case $N \geq 1$.

THEOREM 1.5 ([1]) <u>Let</u> f <u>verify</u> (1.4), <u>then</u>

$$sc^-(w-(H^{1,p}(\Omega))^N) \int_\Omega f(x,Du) = \int_\Omega g(x,Du)$$

<u>where</u> $g(x,\cdot)$ <u>is the greatest quasiconvex function not greater than</u> $f(x,\cdot)$.(*)

This kind of result, in some particular cases, has been recently obtained, with a different method, by [10] .

Finally let us mention another type of problem which has been solved only in the case N=p=1.

We mean the problem of the representation of $sc^-(L^1)F$ of the functional

$$F(u) = \int_\Omega f(x,u,Du),$$

outside of the "continuity space" $H^{1,1}$, e.g. for $u \in BV$.

For p=N=1 the first results were obtained by SERRIN [33] and GOFFMAN-SERRIN [23] in the case

$$F(u) = \int_\Omega f(Du) .$$

Several extension have been given of these results : [6], [17], [11] [22], [7], the most general beeing those of [11].

(*) Actually theorems 1.4 and 1.5 are particular cases of the results of [1] , where $f = f(x,y,z)$, $x \in R^n, y \in R^N$, $z \in R^{nN}$.

2. SOME REGULARITY RESULTS FOR LOCAL MINIMA OF FUNCTIONALS

Let $\mathcal{A}(\Omega)$ be the family of all the open subsets of a fixed open set Ω of R^n and $F: \mathcal{A}(\Omega) \times (H^{1,p}_{loc}(\Omega))^N \longrightarrow R$ be a functional satisfying the condition:

$$(2.1) \qquad \int_A |Du|^p \leq F(A,u) \leq \int_A (a(x) + b|u|^p + c|Du|^p)$$

for $A \in \mathcal{A}(\Omega)$ and $u \in (H^{1,p}_{loc}(\Omega))^N$.

In this section we want to present a Caccioppoli type inequality for a local minimum u of F on Ω, that is for $u \in (H^{1,p}_{loc}(\Omega))^N$ such that $\forall A \in \mathcal{A}(\Omega)$:

$$(2.2) \qquad F(A,u) \leq F(A,w) \qquad \forall w \in u + (H^{1,p}_0(A))^N.$$

From Caccioppoli type inequality we can derive other regularity theorems, generalizing some recent results relative to the case that F is a multiple integral of the Calculus of Variations [2], [20], [27], without Euler equation.

In particular, in the case p=n, we shall obtain a Liouville theorem, generalizing the classical one relative to partial differential equations [30], [34], and quasiconformal mappings, [31]

THEOREM 2.1 (Caccioppoli type) *Let Ω be an open set in R^n, u a local minimum of F, a functional satisfying (2.1).*

Let $x_0 \in \Omega$, $0<t<s<\text{dist}(x_0, \partial\Omega)$, $\nu \geq 1$ an integer. Then for any $v \in (H^{1,p}_{loc}(\Omega))^N$ we have:

$$(2.3) \qquad \int_{B_t} |Du|^p \leq \frac{2^{p-1}c}{\nu} \int_{B_s - B_t} |Du|^p + \frac{2^{p-1}c}{\nu} \frac{(\nu+1)^p}{(s-t)^p} \int_{B_s - B_t} |v-u|^p +$$

$$+ \int_{B_s} [a(x) + b(|v|^p + |u|^p) + 2^{p-1} c\, |Dv|^p].$$

PROOF ([32]).

Let us now prove a simple algebraic lemma.

LEMMA 2.1 *Let* $g: R^+ \longrightarrow R^+$ *be an increasing function such that*

(2.4) $$t<s \implies g(t) \le \theta g(s) + \frac{A\, s^n}{(s-t)^n}$$

where $A>0$ *and* $0<\theta<1$. *Then* $g \in L^\infty(R^+)$.

PROOF. Let $r>0$, $0<\tau<1$ to be chosen, and define a sequence (t_h) in the following manner ([20]):

$$t_o = r, \quad t_{i+1} - t_i = (1-\tau)\tau^i r.$$

By (2.4) we deduce, since $t_{i+1} \le 2r$:

(2.5) $$g(t_i) \le \theta\, g(t_{i+1}) + \frac{2^n A}{(1-\tau)^n \tau^{in}}$$

By iteration we find, using the monotonicity of g,

$$g(r) \le \theta^k g(2r) + \frac{2^n A}{(1-\tau)^n} \sum_{i=0}^{k} \left(\frac{\theta}{\tau^n}\right)^i \qquad \forall\, k.$$

Let us now choose $\theta \cdot \tau^{-n} < 1$, and pass to the limit as $k \to \infty$, then:

$$g(r) \le \frac{2^n A}{(1-\tau)^n} \; \frac{1}{1-\theta\tau^{-n}},$$

and so we have the result.

By theo. 2.1 and lemma 2.1 we deduce the following Liouville type theorem.

THEOREM 2.2 Let $p=n$, $a(x)=b=0$ in (2.1) and assume that $u \in (L^\infty(R^n))^N$ is a local minimum of F on R^n. Then $u(x) = const$.

PROOF. Let us apply (2.3) with $v = 0$. Then we find

$$\int_{B_t} |Du|^n \leq \frac{c'}{\nu}\int_{B_s - B_t} |Du|^n + \frac{c'}{(s-t)^n}\frac{(\nu+1)^n}{\nu}\int_{B_s} |u|^n$$

$$\leq \frac{c'}{\nu}\int_{B_s - B_t} |Du|^n + c'' \frac{(\nu+1)^n}{\nu}\frac{s^n}{(s-t)^n} \|u\|_{(L^\infty)^N}.$$

Set

$$g(t) = \int_{B_t} |Du|^n$$

and apply (for $\nu > c'$) lemma 2.1; then $g(r) \leq L \;\forall\, r>0$. Let us now use again (2.3) with $s= 2t$ and

$$v = \fint_{B_{2t}-B_t} u \, dx = u_t .$$

We have

$$g(t) \leq \frac{c'}{\nu}(g(2t) - g(t)) + \frac{c'}{t^n}\frac{(\nu+1)^n}{\nu}\int_{B_{2t}-B_t} |u-u_t|^n$$

and, by Poincaré inequality

$$\frac{1}{t^n}\int_{B_{2t}-B_t} |u-u_t|^n \leq k \int_{B_{2t}-B_t} |Du|^n$$

we deduce

$$g(t) \leq K(\nu)\,(g(2t)-g(t)),$$

that is:

$$g(t) \leq \frac{K(\nu)}{K(\nu)+1}\, g(2t)$$

By iteration, since $g(t) \leq L \;\forall t$, we find, for any integer $m>1$

$$g(t) \leq \left(\frac{K(\nu)}{K(\nu)+1}\right)^m g(2^m t) \leq \left(\frac{K(\nu)}{K(\nu)+1}\right)^m L$$

for any $t>0$; and so, to the limit as $m \to \infty$, we get

$$g(t)=0 \qquad \forall\, t>0$$

that is $u(x) = \text{const}$.

Another consequence of Caccioppoli inequality is related to the theory of Γ-convergence ([14]).

Let us consider a sequence $F_h : \mathcal{A}(\Omega) \times (H^{1,p}_{loc}(\Omega))^N \longrightarrow R$ of functionals satisfying, for $p>1$, (2.1) uniformly with respect to h.

Then we shall say that $F : \mathcal{A}(\Omega) \times (H^{1,p}_{loc}(\Omega))^N \longrightarrow R$ is the $\Gamma^-((L^p(\Omega))^N)$limit of F_h ([14]) iff $\forall\, A \in \mathcal{A}(\Omega)$:

(i) $u_h \rightarrow u$ in $L^p(A)^N \implies F(A,u) \leq \liminf_h F_h(A,u_h)$

(ii) $\forall u \;\exists w_h \longrightarrow u$ in $L^p(A)^N$ such that $F(A,u) = \lim_h F_h(A,w_h)$.

The following theorem derives from theo 2.1.

THEOREM ([3]). <u>Let</u> F_h, F <u>verify</u> (2.1) <u>uniformly in</u> h. <u>Assume that</u> (i), (ii) <u>hold and that</u>

$u_h \in (H^{1,p}_{loc}(\Omega))^N$ <u>is a local minimum of</u> F_h <u>on</u> Ω

$u_h \longrightarrow u$ <u>in</u> $L^p_{loc}(\Omega)^N$.

<u>Then</u>:

$u_h \rightharpoonup u$ <u>in</u> $(H^{1,p}_{loc}(\Omega))^N$

$u \in (H^{1,p}_{loc}(\Omega))^N$ <u>is a local minimum of</u> F <u>on</u> Ω

$F_h(A,u_h) \longrightarrow F(A,u) \qquad \forall A \subset\subset \Omega$.

This theorem was proved in [2] in the particular case

$$F_h(A,u) = \int_A f_h(x,Du) \qquad N=1$$

$$f_h(x,\cdot) \text{ convex on } R^n$$

and in a more general fashion in [12] , with a completely different proof.

3. EXISTENCE VIA REGULARITY RESULTS.

Let $F = F(A,u)$ be a real functional satisfying (2.1) for $A \in \mathcal{A}(\Omega)$, $u \in (H^{1,p}_{loc}(\Omega))^N$, and moreover the following conditions:

(3.1) $\quad F(\cdot,u)$ trace of a measure on $\mathcal{A}(\Omega)$

(3.2) $\quad u_{/A} = v_{/A} \implies F(A,u) = F(A,v)$

(3.3) $\quad F(A,\cdot)$ strongly lower semicontinuous in $(H^{1,1}(A))^N$.

The main result of this section is the following existence theorem of minima of variational problems of Dirichlet type, which generalizes a result of [17].

THEOREM 3.1 *Let us fix Ω bounded open set in R^n, $u_o \in (H^{1,\infty}(\Omega))^N$, F a functional verifying* (2.1) *with* $p>1$, $a \in L^{1+\sigma}(\Omega)$, (3.1), (3.2) *and* (3.3). *Then there exists a minimizing sequence* (u_h) *of* $F(\Omega,\cdot)$ *on* $u_o+(H^{1,p}_o(\Omega))^N$, *such that for a certain* $\varepsilon>0$, $\forall h \geq h_o$:

$$\|u_h\|_{(H^{1,p+\varepsilon}_{loc}(\Omega))^N} \leq \text{const.}$$

PROOF. As in [17] we choose $v_h \in u_o + (H^{1,1}_o(\Omega))^N$ such that

$$F(\Omega,v_h) \leq \operatorname*{Inf}_{v \in u_o+(H^{1,1}_o(\Omega))^N} F(\Omega,v) + 1/h$$

and then, by Ekeland's theorem [15] we find $u_h \in u_o+(H^{1,p}_o(\Omega))^N = V$ $\|u_h\|_V \leq$ const. , such that $\forall\ A \in \mathcal{A}(\Omega)$:

(3.4) $$F(A,u_h) \leq F(A,v) + \frac{1}{h}\|Du_h-Dv\|_{(L^1(A))^{nN}}$$

for any $v \in u_h+(H^{1,p}_o(A))^N$.

At this point we can write a Caccioppoli inequality (theo.2.1):

for any $x_o \in \Omega$, $R < \frac{1}{2}\text{dist}(x_o, \partial\Omega)$, for any h, ν integers

$$(3.5) \qquad \frac{1}{2^n} \int_{B_R} |Du_h|^p \leq \int_{B_{2R}} (a(x)+2b|u_h|^p + |Du_h|) +$$

$$+ \frac{c}{\nu} \int_{B_{2R}} |Du_h|^p + \frac{\nu+1}{R} \int_{B_{2R}} |u_h - (u_h)_{2R}| +$$

$$\frac{cp^p(\nu+1)^p}{R^p} \int_{B_{2R}} |u_h - (u_h)_{2R}|^p$$

And so, as in [17], set $\mu = \max\{1, \frac{np}{n+p}\}$ and

$$k_h(x) = 2^n(a(x) + 2b|u_h|^p + |Du_h| + 1)^{\mu/p}.$$

By (3.5) and Sobolev imbedding theorem, we find

$$\int_{B_R} |Du_h|^p \leq \beta \left[\left(\int_{B_{2R}} |Du_h|^\mu \right)^{p/\mu} + \int_{B_{2R}} k_h^{p/\mu} \right] +$$

$$+ \frac{2^n c}{\nu} \int_{B_{2R}} |Du_h|^p$$

with $\beta = \beta(n, N, p, c, \ )$.

We can now use Gehring-Giaquinta-Modica lemma [19],[21] and so we find $\varepsilon = \varepsilon(p, n, N, c, b)$ such that for h large

$$\|u_h\|_{(H^{1,p+\varepsilon}_{loc}(\Omega))^N} \leq \gamma \left(\|a\|_{L^{1+\sigma}(\Omega)} + \|u_h\|_{(H^{1,p}(\Omega))^N} \right.$$

with $\gamma = \gamma(N, n, c, b)$.

REFERENCES

[1] E.ACERBI-N.FUSCO, Semicontinuity problems in the Calculus of Variations, Arch. Rational Mech. Anal.(to appear).

[2] H.ATTOUCH-C.SBORDONE, Asymptotic limits for perturbed functionals of Calculus of Variations, Ricerche di Mat., XXIX (1), 85-124 (1980).

[3] H.ATTOUCH-C.SBORDONE, A general homogenization formula for functionals of Calculus of Variations (to appear).

[4] J.M.BALL, Convexity conditions and existence theorems in nonlinear elasticity, Arch. Rational Mech. Anal. 63, 337-403 (1977).

[5] L.D.BERKOWITZ, Lower semicontinuity of integral functionals, Trans. Am. Math. Soc. 192, 51-57 (1974).

[6] M.BONI, Su una definizione dell'integrale multiplo del Calcolo delle Variazioni, Atti Sem. Mat. Fis. Univ. Modena, XIX, (1), 86-106 (1970).

[7] G.BUTTAZZO-G.DAL MASO, Integral representation on $W^{1,\alpha}(\Omega)$ and $BV(\Omega)$ of limits of variational integrals, Rend. Acc. Naz. Lincei LXVI 5, 338-343 (1979).

[8] L.CARBONE-C.SBORDONE, Some properties of Γ-limits of integral functionals, Annali Mat. Pura Appl. IV 122 1-60 (1979)

[9] L.CESARI, Lower semicontinuity and lower closure theorems without seminormality condition, Annali Mat. Pura Appl. 98, 381-397 (1974).

[10] B.DACOROGNA, A relaxation theorem and its application to the equilibrium of gases,Arch. Rational Mech. Anal. (to appear).

[11] G.DAL MASO,Integral representation on $BV(\Omega)$ of Γ-limits of Variational Integrals,Manuscripta Math. 30, 387-416 (1980)

[12] G.DAL MASO-L.MODICA, A general theory of Variational Functionals (to appear)

[13] E.DE GIORGI, Teoremi di semicontinuità nel Calcolo delle Variazioni, Ist.Naz. Alta Mat., Roma (1968-69)

[14] E.DE GIORGI, Convergence problems for functionals and operators, Proc. Int. Meeting "Recent Methods in Nonlinear Analysis" Ed. E. De Giorgi-E.Magenes -U.Mosco, Pitagora ,Bologna (1979).

[15] I.EKELAND, Nonconvex minimization problems, Bull. Amer. Math. Soc. 1, (3), 443-474 (1979).

[16] I.EKELAND-R.TEMAM, Convex Analysis and Variational Problems, North Holland (1976).

[17] F.FERRO, Integral characterization of functionals defined on spaces of BV functions, Rend. Sem. Mat. Univ. Padova

[18] N.FUSCO-G.MOSCARIELLO, L^2-lower semicontinuity of functionals of quadratic type, Annali Mat. Pura Appl, (to appear).

[19] F.W.GEHRING, The L^pintegrability of the partial derivatives of a quasiconformal mapping, Acta Mth. 130, 265-277 (1973).

[20] M.GIAQUINTA-E.GIUSTI, On the regularity of the minima of variational integrals, Acta Math.(to appear).

[21] M.GIAQUINTA-G.MODICA, Regularity results for some classes of higher order non linear elliptic systems, J. fur Reine u. Angew. Math. 311/312,145-169 (1979)

[22] M.GIAQUINTA,G.MODICA,J.SOUCEK, Functionals with linear growth in the Calculus of Variations, Comment. Math. Univ. Carolinae 20, 143-156 (1979)

[23] C.GOFFMAN-J.SERRIN, Sublinear functions of measures and variational integrals, Duke Math. J. 31, 159-178 (1964).

[24] A.IOFFE,On Lower semicontinuity of integral functionals, SIAM J. Cont. Optimization 15, 521-538 (1977).

[25] P.MARCELLINI, Some problems of semicontinuity and Γ-convergence for integrals of the Calculus of Variations, Proc. Int. Meeting "Recent Methods in Nonlinear Analysis" Ed. E. De Giorgi,E.Magenes,U.Mosco, Pitagora, Bologna (1979).

[26] P.MARCELLINI-C.SBORDONE, Semicontinuity problems in the Calculus of Variations, Nonlinear Analysis TMA, 4 (2) 241-257 (1980)

[27] P.MARCELLINI-C.SBORDONE, On the existence of minima of multiple integrals of the Calculus of Variations, J.Math.Pures Appl. (to appear).

[28] C.B.MORREY, Quasiconvexity and the lower semicontinuity of multiple integrals,Pacific J. Math. 2 25-53 (1952).

[29] C.OLECH, A characterization of L^1-weak lower semicontinuity of integral functionals, Bull. Acad. Pol. Sci. Ser. Sci. Math. Astron. Phys. 25 135-142 (1977).

[30] L.A.PELETIER-J.SERRIN, Gradient bounds and Liouville theorems for quasilinear elliptic equations, Annali Sc. Norm. Sup. Pisa IV 5, 65-104 (1978).

[31] J.G.RESHETNYAK, Mapping with bounded deformation as extremals of Dirichlet type integrals, Siberian Math.J. 9, 487-498 (1968)

[32] C.SBORDONE, Lower semicontinuity and regularity of minima of variational functionals, Nonlinear Partial Diff. Eq. and Their Applications, College de France Seminar, Vol. IV (to appear).

[33] J.SERRIN, On the definition and properties of certain variational integrals, Trans. Amer. Math.Soc.101 139-167 (1961).

[34] J.SERRIN, Liouville Theorems and gradient bounds for quasilinear elliptic systems, Archive Rational Mech. Anal. 66, 295-310 (1977).

[35] K.O.WIDMAN, Holder continuity of solutions of elliptic systems, Manuscripta Math. 5, 299-308 (1971).

Carlo Sbordone
Istituto Matematico "R.Caccioppoli"
Università di Napoli
Via Mezzocannone,8
80134 Napoli (Italy)

A FORMULA FOR THE LEVEL SETS OF EPI-LIMITS AND SOME APPLICATIONS

Roger J-B. Wets
Department of Mathematics
University of Kentucky
Lexington Ky 40506
USA

We give a formula for the level sets of the limit function of a sequence of epi-convergent functions. The result is used to characterize the elements of a sequence whose epi-limit is inf-compact. Finally, we examine the implications of these results for the convergence of the infima and the solution (minimizing) sets. We restrict ourselves to the case when the functions are defined on R^n. However, the presentation is such that, either with the Mosco topology for epi-convergence in the reflexive Banach case, or with the De Giorgi topologies in the more general case, the arguments remain similar to those used here.
We start with a quick review of epi-convergence which at the same time allow us to introduce some notations.

Suppose $\{S^\nu \subset R^n\ ,\ \nu = 1,\ldots\}$ is a sequence of sets.

Its *limits inferior* and *superior* are the sets

$$\liminf_{\nu\to\infty} S^\nu = \{x = \lim_{\nu\to\infty} x^\nu \mid x^\nu \in S^\nu \quad \text{for all } \nu = 1,\ldots\}$$

and

$$\limsup_{\nu\to\infty} S^\nu = \{x = \lim_{k\to\infty} x^k \mid x^k \in S^{\nu_k}, k = 1,\dots \text{ for some } \{\nu_k\} \subset N\}.$$

Thus, $\liminf_{\nu\to\infty} S^\nu$ is the set of limit points of all possible sequences $\{x^\nu, \nu=1,\dots$ with $x^\nu \in S^\nu\}$ and $\limsup_{\nu\to\infty} S^\nu$ is the set of all the cluster points of such sequences. Clearly, we always have that

$$\liminf_{\nu\to\infty} S^\nu \subset \limsup_{\nu\to\infty} S^\nu \quad .$$

The sequence is said to have a *limit*, denoted by $\lim_{\nu\to\infty} S^\nu$, if the inclusion can be replaced by an equality.

Let $\{f^\nu, \nu=1,\dots\}$ be a sequence of functions defined on R^n and with values in $\overline{R}$, the extended reals. The *epi-limits inferior* and *superior* are the functions $(li_e f^\nu)$ and $(ls_e f^\nu)$ whose epigraphs are respectively the limits superior and inferior of the sequence of sets $\{\text{epi } f^\nu, \nu=1,\dots\}$ where epi g denotes the *epigraph* of the function g:

$$\text{epi } g = \{(x,\alpha) \mid g(x) \leq \alpha\} \quad .$$

Simply from the definition, and the above inclusion it follows that

$$li_e f^\nu \leq ls_e f^\nu \quad .$$

The sequence $\{f^\nu, \nu=1,\dots\}$ has an *epi-limit*, denoted by $lm_e f^\nu$, if equality holds, and then

$$lm_e f^\nu = li_e f^\nu = ls_e f^\nu \quad .$$

We then also say that the sequence *epi-converges* to $\mathrm{lm}_e f^\nu$, and we write $f^\nu \to_e (\mathrm{lm}_e f^\nu)$.

Thus a function f is the epi-limit of a sequence $\{f^\nu, \nu=1,\ldots\}$ if

$$\mathrm{ls}_e f^\nu \le f \le \mathrm{li}_e f^\nu .$$

Using the definitions, it is not difficult to see that the second inequality will be satisfied, if for every $x \in R^n$

(i_e) for any subsequence of functions $\{f^{\nu_k}, k=1\ldots\}$ and any sequence $\{x^k, k=1,\ldots\}$ converging to x, we have

$$\liminf_{k\to\infty} f^{\nu_k}(x^k) \ge f(x),$$

and the first inequality, if for every $x \in R^n$

(ii_e) there exists a sequence $\{x^\nu, \nu=1,\ldots\}$ converging to x such that

$$\limsup_{\nu\to\infty} f^\nu(x^\nu) \le f(x).$$

For any decreasing sequence of subsets $\{S^\nu, \nu=1,\ldots\}$ of R^n we have that $\lim_{\nu\to\infty} S^\nu$ exists and is given by the formula

$$\lim_{\nu\to\infty} S^\nu = \bigcap_{\nu=1} \mathrm{cl}\ S^\nu$$

Similarly, if the $\{f^\nu : R^n \to \overline{R}, \nu=1,\ldots\}$ is an increasing sequence of functions, i.e., $f^\nu \le f^{\nu+1}$, then the epi-limit exists and is given by

$$\mathrm{lm}_e f^\nu(x) = \lim_{\nu\to\infty} \mathrm{cl}\ f^\nu(x)$$

where cl g is the *lower semicontinuous closure* of g, or equivalently cl g is the function such that epi cl g = cl epi g.

The next theorem gives a characterization of the level sets of the limit function in terms of the level sets of the functions f^ν. For $\alpha \in R$, the *α-level set* of a function g is the set defined by

$$\mathrm{lev}_\alpha g = \{(x,\alpha) \mid g(x) \le \alpha\}.$$

In general, if $f = \lim_{\nu\to\infty} f^\nu$, it does not imply that $\mathrm{lev}_\alpha f = \lim_{\nu\to\infty} \mathrm{lev}_\alpha f^\nu$. Simply think of the decreasing collection of functions

$$f^\nu(x) = \nu^{-1}x^2 \qquad , \ \nu = 1,\ldots$$

that epi-converge to $f \equiv 0$. The $\mathrm{lev}_0 f^\nu = \{0\}$ for all ν, and thus $\lim_{\nu\to\infty} \mathrm{lev}_0 f^\nu = \{0\}$ but $\mathrm{lev}_0 f = R$. It is even possible for the f^ν to epi-converge to f but for some $\alpha \in R$, $\lim_{\nu\to\infty} \mathrm{lev}_\alpha f^\nu$ may not even exist which means that $\liminf_{\nu\to\infty} \mathrm{lev}_\alpha f^\nu$ is strictly included in $\limsup_{\nu\to\infty} \mathrm{lev}_\alpha f^\nu$. Again take $f^\nu(x) = \nu^{-1}x^2$ for all even ν, and $f^\nu \equiv 0$ for all odd indices ν. Then the f^ν epi-converge to $f \equiv 0$. Clearly

$$\begin{aligned} \mathrm{lev}_0 f^\nu &= \{0\} \qquad \text{if } \nu \text{ is odd} \\ &= R \qquad \text{if } \nu \text{ is even} \end{aligned}$$

and thus $\liminf_{\nu\to\infty} \mathrm{lev}_0 f^\nu = \{0\} \neq R = \limsup_{\nu\to\infty} \mathrm{lev}_0 f^\nu$.

1. THEOREM *Suppose* $\{f^\nu = R^n \to \overline{R},\ \nu = 1,\ldots\}$ *is a sequence of functions. Then for all* $\alpha \in R$,

(2) $$\lim_{\alpha'\downarrow\alpha} \limsup_{\nu\to\infty} (\mathrm{lev}_{\alpha'} f^\nu) \subset \mathrm{lev}_\alpha (\mathrm{li}_e f^\nu)$$

and

(3) $$\mathrm{lev}_\alpha(\mathrm{ls}_e f^\nu) \subset \lim_{\alpha'\downarrow\alpha} \liminf_{\nu\to\infty} (\mathrm{lev}_{\alpha'} f^\nu)$$

PROOF. Let $T_{\alpha'} = \limsup_{\nu\to\infty} \text{lev}_{\alpha'} f^\nu$ and $T = \lim_{\alpha'\downarrow\alpha} T_{\alpha'}$. Since the level sets (of any function) are decreasing as $\alpha'\downarrow\alpha$, it follows that the $T_{\alpha'}$ are decreasing as $\alpha'\downarrow\alpha$ and thus

$$T = \lim_{\alpha'\downarrow\alpha} T_{\alpha'} = \bigcap_{\alpha'>\alpha} T_{\alpha'},$$

the sets $T_{\alpha'}$ being closed, as follows directly from the definition of limit superior. It follows that $x \in T$ if and only if $x \in T_{\alpha'}$ for all $\alpha'>\alpha$. The inclusion (2) is trivially satisfied if T is empty. Henceforth, let us assume that T is nonempty. If $x \in T_{\alpha'}$, the definition of limit superior for sequences of sets implies that there necessarily exists a subsequence of functions $\{f^{\nu_k}, k=1,\ldots\}$ and a sequence $\{x^k, k=1,\ldots\}$ converging to x such that for all $k=1,\ldots$

$$x^k \in \text{lev}_{\alpha'} f^{\nu_k}$$

or equivalently such that for all $k=1,\ldots$

$$(x^k,\alpha') \in \text{epi } f^{\nu_k}.$$

Since $\text{epi }(\text{li}_e f^\nu) = \limsup_{\nu\to\infty} \text{epi } f^{\nu_k}$ it follows that $(x,\alpha') = \lim_{k\to\infty}(x^{\nu_k},\alpha') \in \text{epi }(\text{li}_e f^\nu)$ and thus $x \in \text{lev}_{\alpha'}(\text{li}_e f^\nu)$. Hence if $x \in T_{\alpha'}$ for all $\alpha'>\alpha$ it follows that $x \in \text{lev}_{\alpha'}(\text{li}_e f^\nu)$ for all $\alpha'>\alpha$ which implies that $x \in \text{lev}_\alpha(\text{li}_e f^\nu)$ since for any function g $\text{lev}_\alpha g = \bigcap_{\alpha'>\alpha} \text{lev}_{\alpha'} g$.

Let $S_{\alpha'} = \liminf_{\nu\to\infty} \text{lev}_{\alpha'} f^\nu$ and $S = \lim_{\alpha'\downarrow\alpha} S_{\alpha'} = \bigcap_{\alpha'>\alpha} S_{\alpha'}$. Again the inclusion (3) is trivial if $\text{lev}_\alpha(\text{ls}_e f^\nu) = \emptyset$, there only remains to consider the case when $\text{lev}_\alpha(\text{ls}_e f^\nu)$ is nonempty. If $x \in \text{lev}_\alpha(\text{ls}_e f^\nu)$ it implies that there exist (x^ν,α^ν) converging to (x,α) such that

$$(x^\nu,\alpha^\nu) \in \text{epi } f^\nu$$

since by definition $\text{epi}(\text{ls}_e f^\nu) = \liminf_{\nu\to\infty} \text{epi}\, f^\nu$. Since $\alpha = \lim_{\nu\to\infty}\alpha^\nu$, to any $\alpha' > \alpha$ there corresponds ν' such that $\alpha^\nu \leq \alpha'$ for all $\nu \geq \nu'$. This implies that $x^\nu \in \text{lev}_{\alpha'} f^\nu$ for all $\nu \geq \nu'$ and consequently $x \in S_{\alpha'}$. The above holds for every $\alpha' > \alpha$ from which it follows that $x \in S$. This yields the inclusion (3). □

. COROLLARY. *Suppose* $\{f; f^\nu, \nu = 1,\dots\}$ *is a collection of functions defined on* R^n, *with values in the extended reals* $\bar{R}$, *and such that* $f = \text{lm}_e f^\nu$. *Then for all* $\alpha \in R$

$$\text{(5)}\qquad \text{lev}_\alpha f = \lim_{\alpha'\downarrow\alpha} \limsup_{\nu\to\infty} (\text{lev}_{\alpha'} f^\nu)$$

$$= \lim_{\alpha'\downarrow\alpha} \liminf_{\nu\to\infty} (\text{lev}_{\alpha'} f^\nu) \quad .$$

PROOF. Since $f = \text{lm}_e f^\nu = \text{li}_e f^\nu = \text{ls}_e f^\nu$, it follows from the theorem that

$$\lim_{\alpha'\downarrow\alpha} \limsup_{\nu\to\infty} (\text{lev}_{\alpha'} f^\nu) \subset \text{lev}_\alpha f \subset \lim_{\alpha'\downarrow\alpha} \liminf_{\nu\to\infty} (\text{lev}_{\alpha'} f^\nu)$$

The relations (5) now simply follow from the fact that for any α', $\liminf_{\nu\to\infty} (\text{lev}_{\alpha'} f^\nu) \subset \limsup_{\nu\to\infty} (\text{lev}_{\alpha'} f^\nu)$. □

Equipped with his formulas, we now turn to the characterization of the elements of a sequence of functions $\{f^\nu, \nu = 1,\dots\}$ whose epi-limit (exists and) is inf-compact. The first couple of propositions are proved in [1].

. PROPOSITION. *Suppose* $\{S^\nu, \nu = 1\dots\}$ *is a consequence of subsets of* R^n. *Then* $\limsup_{\nu\to\infty} S^\nu = \emptyset$, *or equivalently* $\lim_{\nu\to\infty} S^\nu = \emptyset$, *if and only if to any bounded set* D *there corresponds an index* ν_D *such that*

$$S^\nu \cap D = \emptyset \textit{ for all } \nu \geq \nu_D \quad .$$

. PROPOSITION. *Suppose* S *and* $\{S^\nu, \nu = 1,\dots\}$ are *subsets of* R^n *with* S *closed. Then*

$$S \subset \liminf_{\nu\to\infty} S^\nu \text{ if and only if for all } \varepsilon > 0,\ \lim_{\nu\to\infty} S\setminus \varepsilon^\circ S^\nu = \emptyset,$$

and

$$S \supset \limsup_{\nu\to\infty} S^\nu \text{ if and only if for all } \varepsilon > 0,\ \lim_{\nu\to\infty} S^\nu\setminus \varepsilon^\circ S = \emptyset.$$

where

$\varepsilon^\circ D$ *denotes the (open)* ε-enlargement *of the set* D, *i.e.*

$$\varepsilon^\circ D = \{x \in R^n \mid \text{dist}(x,D) < \varepsilon\} \quad .$$

The next proposition improves somewhat a result of [2] concerning the convergence of connected sets.

8. PROPOSITION. *Suppose* $\{S^\nu, \nu = 1,\ldots\}$ *is a sequence of connected subsets of* R^n *such that* $\limsup_{\nu\to\infty} S^\nu$ *is bounded. Then there exists* ν' *such that for* $\nu \geq \nu'$, *the sets* S^ν *are uniformly bounded.*

PROOF. Let $S = \limsup_{\nu\to\infty} S^\nu$. For all $\varepsilon > 0$, we have that

$$S^\nu = (S^\nu\setminus \varepsilon^\circ S) \cup (S^\nu \cap \varepsilon^\circ S) \quad .$$

From Proposition 7, it follows $\lim_{\nu\to\infty}(S^\nu\setminus\varepsilon^\circ S) = \emptyset$. In view of Proposition 6, this implies that for any $\beta > \varepsilon$,

$$(S^\nu\setminus \varepsilon^\circ S) \cap \beta^\circ S = \emptyset$$

for all ν sufficiently; recall that S is bounded by assumption and thus so is $\beta^\circ S$. Hence for ν sufficiently large $S^\nu \subset \varepsilon^\circ S$ since otherwise the sets S^ν would have to be disconnected since we could write $S^\nu = (S^\nu \cap \varepsilon^\circ S) \cup (S^\nu\setminus\beta^\circ S)$ with $\beta > \varepsilon$. □

9. THEOREM. *Suppose* $\{f^\nu : R^n \to \overline{R},\ \nu = 1,\ldots\}$ *is a sequence of lower semicontinuous functions with connected level sets and such that the epi-limit inferior* $\mathrm{li}_e f^\nu$ *is inf-compact. Then the functions* f_ν *are uniformly inf-compact, in the sense that for all* α *there exists* ν_α *such that for all* $\nu \geq \nu_\alpha$, *the level sets* $\mathrm{lev}_\alpha f^\nu$ *are uniformly compact.*

PROOF. We first note that for all $\alpha \in R$, we have

$$\limsup_{\nu\to\infty} \mathrm{lev}_\alpha \subset \lim_{\alpha'\downarrow\alpha} \limsup_{\nu\to\infty} \mathrm{lev}_{\alpha'} f^\nu .$$

The inclusion is certainly true if $\limsup_{\nu\to\infty} \mathrm{lev}_\alpha f^\nu$ is empty. Otherwise $x \in \limsup_{\nu\to\infty} \mathrm{lev}_\alpha f^\nu$ implies that there exists a subsequence $\{\nu_k, k = 1,\dots\}$ and $\{x^k, k = 1,\dots\}$ a sequence converging to x such that $x^k \in \mathrm{lev}_\alpha f^{\nu_k}$ for all $\alpha' > \alpha$. Hence $x \in \lim_{\alpha'\downarrow\alpha} \limsup_{\nu\to\infty} \mathrm{lev}_{\alpha'} f^\nu$ which completes the proof of the inclusion.

We now combine the above with (2) to obtain

$$\limsup_{\nu\to\infty} \mathrm{lev}_\alpha f^\nu \subset \mathrm{lev}_\alpha (\mathrm{li}_e f^\nu) .$$

By assumption for all α, $\mathrm{lev}_\alpha(\mathrm{li}_e f^\nu)$ is compact. A straightforward application of Proposition 8 completes the proof, recalling that for all ν the $\mathrm{lev}_\alpha f^\nu$ are closed since the functions f^ν are lower semicontinuous. □

10. COROLLARY. *Suppose* $\{f^\nu : R^n \to \bar{R}, \nu = 1,\dots\}$ *is a sequence of lower semicontinuous functions with connected level sets, that epi-converges to* f. *Then* f *is inf-compact if and only if the* f^ν *are uniformly inf-compact.*

PROOF. If the f^ν epi-converge to f, then $\mathrm{li}_e f^\nu = f$ and thus the only if part follows from the Theorem. The if part follows from (5). The uniform inf-compactness of the f^ν implies that the $\{S_{\alpha'} = \liminf_{\nu\to\infty} \mathrm{lev}_{\alpha'} f^\nu, \alpha' > \alpha\}$ form a decreasing sequence of compact sets as $\alpha' \downarrow \alpha$ and thus $\mathrm{lev}_\alpha f = \lim_{\alpha'\downarrow\alpha} S_{\alpha'}$ is compact. □

11. COROLLARY. *Suppose* $\{f^\nu : R^n \to \bar{R}, \nu = 1,\dots\}$ *is a sequence of lower semicontinuous convex functions that epi-converges to the (necessarily lower semicontinuous and convex) function* f. *Then* f *is inf-compact if and only if the* f^ν *are uniformly inf-compact.*

PROOF. The level sets of convex functions are convex and thus connected. □

Inf-compactness is usually used to prove the existence of a minimum. It is well-known that a number of weaker conditions can actually be used to arrive at existence. An easy generalization is *quasi-inf-compactness.* A function f is *quasi-inf-compact* if there exists $\alpha \in R$ such that $\text{lev}_\alpha f$ is nonempty and for all $\beta \leq \alpha$, $\text{lev}_\beta f$ is compact. The argument that shows that inf-compact functions have a minimum can also be used in the context of quasi-inf-compact functions. It is not difficult to see how Theorem 9 can be generalized to the case when $\text{li}_e f^\nu$ is quasi-inf-compact. All of this, just to point out that the subsequent results about convergence of infima are not necessarily the sharpest one could possibly obtain by relying on the preceding arguments and results. Thus the next propositions are meant to be illustrative (rather than exhaustive).

12. PROPOSITION. *Suppose* $\{f^\nu : R^n \to \overline{R}, \nu = 1,\ldots\}$ *is a sequence of functions uniformly inf-compact that epi-converges to* f. *Then*

$$\lim_{\nu\to\infty}(\inf f^\nu) = \inf f. \tag{13}$$

PROOF. The inequality

$$\limsup_{\nu\to\infty} (\inf f^\nu) \leq \inf f$$

is well-known as it follows directly from epi-convergence in particular condition (ii_e). To see this let us assume (without loss of generality) that $\inf f < \infty$ and that $\{x^k, k = 1,\ldots\}$ is a sequence in R^n such that $\lim_{k\to\infty} f(x^k) = \inf f$. From ($\text{ii}_e$) it follows that to every x^k there corresponds a sequence $\{x^{k\nu}, \nu = 1,\ldots\}$ converging to x^k such that for all k

$$\limsup_{\nu\to\infty} f^\nu(x^{k\nu}) \leq f(x^k)$$

Since $\inf f^\nu \leq f^\nu(x^{k\nu})$, for all k it follows that

$$\limsup_{\nu\to\infty} (\inf f^\nu) \leq f(x^k)$$

Taking limits on both sides, with respect to k yields the desired relation.

There remains to show that

$$\liminf_{\nu\to\infty} (\inf f^{\nu}) \geq \inf f$$

There is nothing to prove if $\inf f = -\infty$, so we shall only deal with the case when $\inf f > -\infty$. We restrict our attention to the subsequence of indices for which the $\inf f^{\nu}$ converge to $\liminf_{\nu\to\infty} (\inf f^{\nu})$, say

$$\lim_{k\to\infty} (\inf f^{\nu_k}) = \liminf_{\nu\to\infty} (\inf f^{\nu}) \quad .$$

Now, the f^{ν_k} are inf-compact and thus their infima are attained. Let $\{y^k, k = 1,\ldots\}$ be a sequence of points such that for all k, $f^{\nu_k}(y^k) = \inf f^{\nu_k}$. The sequence $\{y^k, k = 1,\ldots\}$ is bounded. To see this first observe that $\limsup_{\nu\to\infty} (\inf f^{\nu}) \leq \inf f$ implies that for any $\delta > 0$

$$f^{\nu_k}(y^k) = \inf f^{\nu_k} \leq \inf f + \delta$$

for k sufficiently large. Thus for those k, $y^k \in \mathrm{lev}_{\delta+\inf f} f^{\nu_k}$. The uniform inf-compactness of the f^{ν} implies that the compact sets $\mathrm{lev}_{\delta+\inf f} f^{\nu}$ are uniformly bounded. Hence the $\{y^k, k = 1,\ldots\}$ admit a cluster point, say y. It now follows from epi-convergence, in particular condition (ii_e), and the above that

$$\lim_{\nu\to\infty} (\inf f^{\nu_k}) = \lim_{k\to\infty} f^{\nu_k}(y^k) \geq f(y) \geq \inf f \quad ,$$

which completes the proof. □

As corollary to this proposition, we obtain a companion to Theorem 7 of [3] and Theorem 1.7 of [4] which were derived via completely different means.

13. COROLLARY. *Suppose* $\{f^\nu : R^n \to \overline{R}, \nu = 1,\dots\}$ *is a sequence of lower semicontinuous convex functions that epi-converge to the (necessarily lower semicontinuous and convex) function* f. *Moreover suppose that either the* $\{f^\nu, \nu = 1,\dots\}$ *are uniformly inf-compact or* f *is inf-compact. Then*

$$\lim_{\nu\to\infty}(\inf f^\nu) = \inf f \quad .$$

PROOF. When the f^ν are convex, the inf-compactness of f yields the uniform inf-compactness of the f^ν as follows from Corollary 11. We are thus in the setting which allows us to apply the Proposition. □

The assumptions of Proposition 12 are not strong enough to allow us to conclude that the solution sets argmin f^ν converge to argmin f. Indeed consider the situation when the f^ν are defined as follows:

$$f^\nu(x) = \begin{cases} \nu^{-1}[|x| - 1] & \text{if } x \in [-1,1] \quad , \\ +\infty & \text{otherwise.} \end{cases}$$

The f^ν epi-converge to the function

$$f(x) = \begin{cases} 0 & \text{if } x \in [-1,1] \\ +\infty & \text{otherwise,} \end{cases}$$

and satisfy all the hypotheses of Proposition 12, even those of Corollary 13, and indeed the infima converge. But the solution sets, argmin $f^\nu = \{0\}$ for all ν do not converge to argmin $f = [-1,1]$. The same situation prevails even if the inf f^ν converge to inf f from above. For example, let

$$f(x) = \max\,[0\;,\;|x| - 1]$$

and for all $x \in R$,

$$f_\nu(x) = \begin{cases} f(x) & \text{if } \nu \text{ is odd} \\ \max\,[\nu^{-1}x^2, f(x)] & \text{if } \nu \text{ is even.} \end{cases}$$

Then the f^ν epi-converge to f, the infima converge but

$$\liminf_{\nu\to\infty} \operatorname{argmin} f^\nu = \{0\}$$

$$\limsup_{\nu\to\infty} \operatorname{argmin} f^\nu = [-1,1] = \operatorname{argmin} f$$

and thus the limit does not exist.

There does not appear to exist easily verifiable conditions that will guarantee the convergence of the argmin sets. We always have the following, cf. [4] for example.

14. PROPOSITION. *Suppose* $\{f^\nu : R^n \to \overline{R}, \nu = 1,\ldots\}$ *is a sequence of functions that epi-converges to* f. *Then*

(15) $$\limsup_{\nu\to\infty} \operatorname{argmin} f^\nu \subset \operatorname{argmin} f.$$

The preceding example has shown that in general, even in very "regular" situations, one cannot expect the inclusion

$$\operatorname{argmin} f \subset \liminf_{\nu\to\infty} \operatorname{argmin} f^\nu$$

to hold. The simple example that follows has all of the following properties: the functions f^ν are convex, uniformly inf-compact, inf f^ν converges to inf f from above and for all $\alpha \in R$

$$\liminf_{\nu\to\infty} \operatorname{lev}_\alpha f^\nu = \limsup_{\nu\to\infty} \operatorname{lev}_\alpha f^\nu \quad .$$

And nonetheless we still do not have that argmin f is the limit of the argmin f^ν. Again let $f(x) = \max\,[0, |x|-1]$ and for all ν

$$f_\nu(x) = \max\,[\nu^{-1}x^2, f(x)] \quad .$$

It thus appears that the search for characterizations of the points that minimize f, should be mostly in terms of formula (15). In particular one should seek conditions which guarantee that

$\limsup_{\nu\to\infty}$ argmin f^ν is nonempty. Sufficient conditions are provided by the assumptions of Proposition 12 (or Corollary 13) as can be gathered from its proof. Formulas (5) however suggest another direction, namely to replace argmin f^ν by ε-argmin $f^\nu = \{x \in R^n | f^\nu(x) \leq \inf f^\nu + \varepsilon\}$. Indeed this allows us to obtain argmin f as an inferior limit of the ε-argmin f^ν. The proposition below is essentially proven in [5].

16. PROPOSITION. *Suppose* $\{f^\nu : R^n \to \overline{R}, \nu = 1, \ldots\}$ *is a sequence of functions that epi-converge to* f, *and* inf f *is finite. Then*

$$\lim_{\nu\to\infty}(\inf f_\nu) = \inf f$$

if and only if

$$\operatorname{argmin} f = \lim_{\varepsilon\downarrow 0} \liminf_{\nu\to\infty} \ \varepsilon\text{-argmin } f_\nu \quad ,$$

$$= \lim_{\varepsilon\downarrow 0} \limsup_{\nu\to\infty} \ \varepsilon\text{-argmin } f_\nu \quad .$$

REFERENCES

[1] G. Salinetti and R. Wets, On the convergence of closed-valued measurable multifunctions, *Trans.Amer.Math.Soc.* 266(198), 275-289.

[2] G. Salinetti and R. Wets, On the convergence of sequences of convex sets in finite dimensions, *Siam Review* 21(1979), 18-33.

[3] R. Wets, Convergence of convex functions, variational inequalities and convex optimization problems, in *Variational Inequalities and Complementarity Problems*, eds. R. Cottle, F. Giannessi and J-L. Lions. J. Wiley & Sons, New York, 1980. 375-403.

[4] R. Robert, Contributions á l'Analyse Non Linéaire, Thèse, Universite de Grenoble, 1976.

[5] H. Attouch and R. Wets, Approximation and convergence in nonlinear optimization, in *Nonlinear Programming* 4, eds. O. Mangasarian, R. Meyer and S. Robinson, Academic Press, New York, 1981, 367-394.

Supported in part by a Guggenheim Fellowship.

'ol. 817: L. Gerritzen, M. van der Put, Schottky Groups and Mumford :urves. VIII, 317 pages. 1980.

'ol. 818: S. Montgomery, Fixed Rings of Finite Automorphism Groups f Associative Rings. VII, 126 pages. 1980.

'ol. 819: Global Theory of Dynamical Systems. Proceedings, 1979. :dited by Z. Nitecki and C. Robinson. IX, 499 pages. 1980.

'ol. 820: W. Abikoff, The Real Analytic Theory of Teichmüller Space. 'II, 144 pages. 1980.

'ol. 821: Statistique non Paramétrique Asymptotique. Proceedings, 979. Edited by J.-P. Raoult. VII, 175 pages. 1980.

ol. 822: Séminaire Pierre Lelong-Henri Skoda, (Analyse) Années 978/79. Proceedings. Edited by P. Lelong et H. Skoda. VIII, 56 pages, 1980.

ol. 823: J. Král, Integral Operators in Potential Theory. III, 171 ages. 1980.

ol. 824: D. Frank Hsu, Cyclic Neofields and Combinatorial Designs. 'I, 230 pages. 1980.

'ol. 825: Ring Theory, Antwerp 1980. Proceedings. Edited by F. van)ystaeyen. VII, 209 pages. 1980.

'ol. 826: Ph. G. Ciarlet et P. Rabier, Les Equations de von Kármán. 'I, 181 pages. 1980.

ol. 827: Ordinary and Partial Differential Equations. Proceedings, 978. Edited by W. N. Everitt. XVI, 271 pages. 1980.

'ol. 828: Probability Theory on Vector Spaces II. Proceedings, 1979. :dited by A. Weron. XIII, 324 pages. 1980.

'ol. 829: Combinatorial Mathematics VII. Proceedings, 1979. Edited y R. W. Robinson et al.. X, 256 pages. 1980.

'ol. 830: J. A. Green, Polynomial Representations of GL_n. VI, 118 ages. 1980.

ol. 831: Representation Theory I. Proceedings, 1979. Edited by . Dlab and P. Gabriel. XIV, 373 pages. 1980.

'ol. 832: Representation Theory II. Proceedings, 1979. Edited by . Dlab and P. Gabriel. XIV, 673 pages. 1980.

ol. 833: Th. Jeulin, Semi-Martingales et Grossissement d'une iltration. IX, 142 Seiten. 1980.

'ol. 834: Model Theory of Algebra and Arithmetic. Proceedings, 979. Edited by L. Pacholski, J. Wierzejewski, and A. J. Wilkie. VI, 10 pages. 1980.

ol. 835: H. Zieschang, E. Vogt and H.-D. Coldewey, Surfaces and 'lanar Discontinuous Groups. X, 334 pages. 1980.

'ol. 836: Differential Geometrical Methods in Mathematical Physics. 'roceedings, 1979. Edited by P. L. García, A. Pérez-Rendón, and . M. Souriau. XII, 538 pages. 1980.

ol. 837: J. Meixner, F. W. Schäfke and G. Wolf, Mathieu Functions nd Spheroidal Functions and their Mathematical Foundations urther Studies. VII, 126 pages. 1980.

'ol. 838: Global Differential Geometry and Global Analysis. Proeedings 1979. Edited by D. Ferus et al. XI, 299 pages. 1981.

ol. 839: Cabal Seminar 77 - 79. Proceedings. Edited by A. S. echris, D. A. Martin and Y. N. Moschovakis. V, 274 pages. 1981.

ol. 840: D. Henry, Geometric Theory of Semilinear Parabolic Equaons. IV, 348 pages. 1981.

ol. 841: A. Haraux, Nonlinear Evolution Equations- Global Behaviour f Solutions. XII, 313 pages. 1981.

ol. 842: Séminaire Bourbaki vol. 1979/80. Exposés 543-560. /, 317 pages. 1981.

'ol. 843: Functional Analysis, Holomorphy, and Approximation heory. Proceedings. Edited by S. Machado. VI, 636 pages. 1981.

ol. 844: Groupe de Brauer. Proceedings. Edited by M. Kervaire and I. Ojanguren. VII, 274 pages. 1981.

Vol. 845: A. Tannenbaum, Invariance and System Theory: Algebraic and Geometric Aspects. X, 161 pages. 1981.

Vol. 846: Ordinary and Partial Differential Equations, Proceedings. Edited by W. N. Everitt and B. D. Sleeman. XIV, 384 pages. 1981.

Vol. 847: U. Koschorke, Vector Fields and Other Vector Bundle Morphisms - A Singularity Approach. IV, 304 pages. 1981.

Vol. 848: Algebra, Carbondale 1980. Proceedings. Ed. by R. K. Amayo. VI, 298 pages. 1981.

Vol. 849: P. Major, Multiple Wiener-Itô Integrals. VII, 127 pages. 1981.

Vol. 850: Séminaire de Probabilités XV. 1979/80. Avec table générale des exposés de 1966/67 à 1978/79. Edited by J. Azéma and M. Yor. IV, 704 pages. 1981.

Vol. 851: Stochastic Integrals. Proceedings, 1980. Edited by D. Williams. IX, 540 pages. 1981.

Vol. 852: L. Schwartz, Geometry and Probability in Banach Spaces. X, 101 pages. 1981.

Vol. 853: N. Boboc, G. Bucur, A. Cornea, Order and Convexity in Potential Theory: H-Cones. IV, 286 pages. 1981.

Vol. 854: Algebraic K-Theory. Evanston 1980. Proceedings. Edited by E. M. Friedlander and M. R. Stein. V, 517 pages. 1981.

Vol. 855: Semigroups. Proceedings 1978. Edited by H. Jürgensen, M. Petrich and H. J. Weinert. V, 221 pages. 1981.

Vol. 856: R. Lascar, Propagation des Singularités des Solutions d'Equations Pseudo-Différentielles à Caractéristiques de Multiplicités Variables. VIII, 237 pages. 1981.

Vol. 857: M. Miyanishi. Non-complete Algebraic Surfaces. XVIII, 244 pages. 1981.

Vol. 858: E. A. Coddington, H. S. V. de Snoo: Regular Boundary Value Problems Associated with Pairs of Ordinary Differential Expressions. V, 225 pages. 1981.

Vol. 859: Logic Year 1979-80. Proceedings. Edited by M. Lerman, J. Schmerl and R. Soare. VIII, 326 pages. 1981.

Vol. 860: Probability in Banach Spaces III. Proceedings, 1980. Edited by A. Beck. VI, 329 pages. 1981.

Vol. 861: Analytical Methods in Probability Theory. Proceedings 1980. Edited by D. Dugué, E. Lukacs, V. K. Rohatgi. X, 183 pages. 1981.

Vol. 862: Algebraic Geometry. Proceedings 1980. Edited by A. Libgober and P. Wagreich. V, 281 pages. 1981.

Vol. 863: Processus Aléatoires à Deux Indices. Proceedings, 1980. Edited by H. Korezlioglu, G. Mazziotto and J. Szpirglas. V, 274 pages. 1981.

Vol. 864: Complex Analysis and Spectral Theory. Proceedings, 1979/80. Edited by V. P. Havin and N. K. Nikol'skii, VI, 480 pages. 1981.

Vol. 865: R. W. Bruggeman, Fourier Coefficients of Automorphic Forms. III, 201 pages. 1981.

Vol. 866: J.-M. Bismut, Mécanique Aléatoire. XVI, 563 pages. 1981.

Vol. 867: Séminaire d'Algèbre Paul Dubreil et Marie-Paule Malliavin. Proceedings, 1980. Edited by M.-P. Malliavin. V, 476 pages. 1981.

Vol. 868: Surfaces Algébriques. Proceedings 1976-78. Edited by J. Giraud, L. Illusie et M. Raynaud. V, 314 pages. 1981.

Vol. 869: A. V. Zelevinsky, Representations of Finite Classical Groups. IV, 184 pages. 1981.

Vol. 870: Shape Theory and Geometric Topology. Proceedings, 1981. Edited by S. Mardešić and J. Segal. V, 265 pages. 1981.

Vol. 871: Continuous Lattices. Proceedings, 1979. Edited by B. Banaschewski and R.-E. Hoffmann. X, 413 pages. 1981.

Vol. 872: Set Theory and Model Theory. Proceedings, 1979. Edited by R. B. Jensen and A. Prestel. V, 174 pages. 1981.

Vol. 873: Constructive Mathematics, Proceedings, 1980. Edited by F. Richman. VII, 347 pages. 1981.

Vol. 874: Abelian Group Theory. Proceedings, 1981. Edited by R. Göbel and E. Walker. XXI, 447 pages. 1981.

Vol. 875: H. Zieschang, Finite Groups of Mapping Classes of Surfaces. VIII, 340 pages. 1981.

Vol. 876: J. P. Bickel, N. El Karoui and M. Yor. Ecole d'Eté de Probabilités de Saint-Flour IX – 1979. Edited by P. L. Hennequin. XI, 280 pages. 1981.

Vol. 877: J. Erven, B.-J. Falkowski, Low Order Cohomology and Applications. VI, 126 pages. 1981.

Vol. 878: Numerical Solution of Nonlinear Equations. Proceedings, 1980. Edited by E. L. Allgower, K. Glashoff, and H.-O. Peitgen. XIV, 440 pages. 1981.

Vol. 879: V. V. Sazonov, Normal Approximation – Some Recent Advances. VII, 105 pages. 1981.

Vol. 880: Non Commutative Harmonic Analysis and Lie Groups. Proceedings, 1980. Edited by J. Carmona and M. Vergne. IV, 553 pages. 1981.

Vol. 881: R. Lutz, M. Goze, Nonstandard Analysis. XIV, 261 pages. 1981.

Vol. 882: Integral Representations and Applications. Proceedings, 1980. Edited by K. Roggenkamp. XII, 479 pages. 1981.

Vol. 883: Cylindric Set Algebras. By L. Henkin, J. D. Monk, A. Tarski, H. Andréka, and I. Németi. VII, 323 pages. 1981.

Vol. 884: Combinatorial Mathematics VIII. Proceedings, 1980. Edited by K. L. McAvaney. XIII, 359 pages. 1981.

Vol. 885: Combinatorics and Graph Theory. Edited by S. B. Rao. Proceedings, 1980. VII, 500 pages. 1981.

Vol. 886: Fixed Point Theory. Proceedings, 1980. Edited by E. Fadell and G. Fournier. XII, 511 pages. 1981.

Vol. 887: F. van Oystaeyen, A. Verschoren, Non-commutative Algebraic Geometry, VI, 404 pages. 1981.

Vol. 888: Padé Approximation and its Applications. Proceedings, 1980. Edited by M. G. de Bruin and H. van Rossum. VI, 383 pages. 1981.

Vol. 889: J. Bourgain, New Classes of $\mathcal{L}^p$-Spaces. V, 143 pages. 1981.

Vol. 890: Model Theory and Arithmetic. Proceedings, 1979/80. Edited by C. Berline, K. McAloon, and J.-P. Ressayre. VI, 306 pages. 1981.

Vol. 891: Logic Symposia, Hakone, 1979, 1980. Proceedings, 1979, 1980. Edited by G. H. Müller, G. Takeuti, and T. Tugué. XI, 394 pages. 1981.

Vol. 892: H. Cajar, Billingsley Dimension in Probability Spaces. III, 106 pages. 1981.

Vol. 893: Geometries and Groups. Proceedings. Edited by M. Aigner and D. Jungnickel. X, 250 pages. 1981.

Vol. 894: Geometry Symposium. Utrecht 1980, Proceedings. Edited by E. Looijenga, D. Siersma, and F. Takens. V, 153 pages. 1981.

Vol. 895: J.A. Hillman, Alexander Ideals of Links. V, 178 pages. 1981.

Vol. 896: B. Angéniol, Familles de Cycles Algébriques – Schéma de Chow. VI, 140 pages. 1981.

Vol. 897: W. Buchholz, S. Feferman, W. Pohlers, W. Sieg, Iterated Inductive Definitions and Subsystems of Analysis: Recent Proof-Theoretical Studies. V, 383 pages. 1981.

Vol. 898: Dynamical Systems and Turbulence, Warwick, 1980. Proceedings. Edited by D. Rand and L.-S. Young. VI, 390 pages. 1981.

Vol. 899: Analytic Number Theory. Proceedings, 1980. Edited by M.I. Knopp. X, 478 pages. 1981.

Vol. 900: P. Deligne, J. S. Milne, A. Ogus, and K.-Y. Shih, Hodge Cycles, Motives, and Shimura Varieties. V, 414 pages. 1982.

Vol. 901: Séminaire Bourbaki vol. 1980/81 Exposés 561–578. III, 299 pages. 1981.

Vol. 902: F. Dumortier, P.R. Rodrigues, and R. Roussarie, Germs of Diffeomorphisms in the Plane. IV, 197 pages. 1981.

Vol. 903: Representations of Algebras. Proceedings, 1980. Edited by M. Auslander and E. Lluis. XV, 371 pages. 1981.

Vol. 904: K. Donner, Extension of Positive Operators and Korovkin Theorems. XII, 182 pages. 1982.

Vol. 905: Differential Geometric Methods in Mathematical Physics. Proceedings, 1980. Edited by H.-D. Doebner, S.J. Andersson, and H.R. Petry. VI, 309 pages. 1982.

Vol. 906: Séminaire de Théorie du Potentiel, Paris, No. 6. Proceedings. Edité par F. Hirsch et G. Mokobodzki. IV, 328 pages. 1982.

Vol. 907: P. Schenzel, Dualisierende Komplexe in der lokalen Algebra und Buchsbaum-Ringe. VII, 161 Seiten. 1982.

Vol. 908: Harmonic Analysis. Proceedings, 1981. Edited by F. Ricci and G. Weiss. V, 325 pages. 1982.

Vol. 909: Numerical Analysis. Proceedings, 1981. Edited by J.P. Hennart. VII, 247 pages. 1982.

Vol. 910: S.S. Abhyankar, Weighted Expansions for Canonical Desingularization. VII, 236 pages. 1982.

Vol. 911: O.G. Jørsboe, L. Mejlbro, The Carleson-Hunt Theorem on Fourier Series. IV, 123 pages. 1982.

Vol. 912: Numerical Analysis. Proceedings, 1981. Edited by G. A. Watson. XIII, 245 pages. 1982.

Vol. 913: O. Tammi, Extremum Problems for Bounded Univalent Functions II. VI, 168 pages. 1982.

Vol. 914: M. L. Warshauer, The Witt Group of Degree k Maps and Asymmetric Inner Product Spaces. IV, 269 pages. 1982.

Vol. 915: Categorical Aspects of Topology and Analysis. Proceedings, 1981. Edited by B. Banaschewski. XI, 385 pages. 1982.

Vol. 916: K.-U. Grusa, Zweidimensionale, interpolierende Lg-Splines und ihre Anwendungen. VIII, 238 Seiten. 1982.

Vol. 917: Brauer Groups in Ring Theory and Algebraic Geometry. Proceedings, 1981. Edited by F. van Oystaeyen and A. Verschoren. VIII 300 pages. 1982.

Vol. 918: Z. Semadeni, Schauder Bases in Banach Spaces of Continuous Functions. V, 136 pages. 1982.

Vol. 919: Séminaire Pierre Lelong – Henri Skoda (Analyse) Années 1980/81 et Colloque de Wimereux, Mai 1981. Proceedings. Edité par P. Lelong et H. Skoda. VII, 383 pages. 1982.

Vol. 920: Séminaire de Probabilités XVI, 1980/81. Proceedings Edité par J. Azéma et M. Yor. V, 622 pages. 1982.

Vol. 921: Séminaire de Probabilités XVI, 1980/81. Supplément: Géométrie Différentielle Stochastique. Proceedings. Edité par J. Azéma et M. Yor. III, 285 pages. 1982.

Vol. 922: B. Dacorogna, Weak Continuity and Weak Lower Semicontinuity of Non-Linear Functionals. V, 120 pages. 1982.

Vol. 923: Functional Analysis in Markov Processes. Proceedings 1981. Edited by M. Fukushima. V, 307 pages. 1982.

Vol. 924: Séminaire d'Algèbre Paul Dubreil et Marie-Paule Malliavin Proceedings, 1981. Edité par M.-P. Malliavin. V, 461 pages. 1982

Vol. 925: The Riemann Problem, Complete Integrability and Arithmetic Applications. Proceedings, 1979-1980. Edited by D. Chudnovsky and G. Chudnovsky. VI, 373 pages. 1982.

Vol. 926: Geometric Techniques in Gauge Theories. Proceedings 1981. Edited by R. Martini and E.M.de Jager. IX, 219 pages. 1982